Preface

Circadian rhythms align the organism's molecular, cellular, and behavioral activity to deal with cyclic environmental changes due to the earth's rotation around the sun. Many aspects of modern lifestyle contribute to an increased risk of circadian rhythm and sleep disturbances. These disruptions affect essential processes related to immune response and brain function, and the reverse is also true. From chronic diseases to mental health disorders, the roles of circadian misalignment, sleep disturbances, and inflammatory processes are pivotal in understanding the pathophysiology of many diseases. Therefore, it is crucial to clearly understand the intricate relationship between circadian rhythms, sleep, and inflammation.

The field of circadian biology has evolved from investigating the mechanisms of the biological clock to examining its significance in health and disease. These advancements have attracted researchers from diverse disciplines, such as infectious diseases, metabolism, psychiatry, neurodegeneration, and aging. As a result, there has been a growing focus on the interaction between sleep and circadian rhythms and their impact on inflammatory responses and immune function.

This book, "Circadian Rhythms, Sleep, and Inflammation," seeks to connect these interrelated fields by surveying the latest scientific developments. It is designed for researchers, clinicians, and students interested in understanding how disruptions in circadian rhythms and sleep patterns can influence inflammation and, consequently, overall health, as well as how inflammation can, in turn, affect these processes.

We begin with a detailed examination of the circadian regulation of inflammation and then proceed to explore the physiological and pathological aspects of sleep, including its restorative functions and the effects of sleep deprivation. Next, we investigate the interplay between sleep and circadian rhythms, focusing on how these systems impact inflammatory responses and immune function through brain–periphery interactions. The subsequent chapters examine specific health conditions where the interplay between circadian rhythms, sleep, and inflammation is crucial, including aging, neurodegenerative diseases, and psychiatric disorders. By presenting the latest research and clinical insights, we aim to offer a comprehensive understanding of how circadian rhythms, sleep, and inflammation intersect to impact human health. As our knowledge in these areas grows, so does our capacity to develop targeted interventions to address their disruptions. We hope this book will serve as both an educational resource and a catalyst for further research in this rapidly evolving field.

We extend our gratitude to the contributors whose expertise and dedication have made this volume possible. We hope this book will be a valuable resource for those studying and treating conditions related to circadian and sleep disturbances.

To sum up, we invite readers to explore the multifaceted connections between circadian rhythms, sleep, and inflammation and to consider the broader implications for health and disease.

Philadelphia, PA, USA Pawan K. Jha
Heidelberg, Germany Hanan Bouaouda

Contents

Circadian Regulation of Neuroinflammation

Saurav Saha and Aishiki Banerjee

Abstract The circadian clock, an intrinsic 24-h oscillator, is intricately linked to the immune system, playing a pivotal role in regulating various biological processes such as immunological responses, neuroinflammation, glial function, and redox homeostasis. The rhythmic expression patterns of cytokines and their receptors, along with circadian-controlled substances like glucocorticoids, melatonin, and leptin, serve as critical modulators for the immune system's impact on circadian rhythms. Disruption of the circadian clock profoundly affects essential processes underlying immune response and brain function. Despite substantial evidence indicating the mutual influence between the circadian system and neuroinflammation, the precise mechanisms connecting circadian clocks, the immune system, and neuroinflammatory pathways remain elusive. This section delves into the interplay between the circadian system and neuroinflammation, providing a comprehensive overview of circadian regulation in both central and peripheral immune cells, with a particular focus on those implicated in neuroinflammatory processes. Understanding these complex interactions is crucial for identifying novel therapeutic targets and developing early interventions for neurodegenerative disorders, potentially revolutionizing treatment approaches by elucidating the mechanistic links between the circadian clock, inflammation, and neurodegeneration.

Keywords Biological clock · Circadian rhythm · Immune system · Inflammation Inflammatory mediators · Neuroinflammatory diseases

S. Saha (✉)
Division of Geriatrics, Department of Internal Medicine, Kansas University Medical Center, Kansas City, KS, USA

Department of Physiology and Aging, University of Florida, Gainesville, FL, USA

A. Banerjee
Department of Microbiology and Cell Science, University of Florida, Gainesville, FL, USA

P. K. Jha, H. Bouaouda (eds.), *Circadian Rhythms, Sleep and Inflammation*, Progress in Inflammation Research 93,
https://doi.org/10.1007/978-3-031-74744-1_1

1 Introduction

The biological clock, which regulates biological activities to adjust to the regular 24-h light/dark cycles caused by the Earth's rotation, is found in almost all species [1]. Numerous physiological and behavioral processes in mammals, such as the sleep-wake cycle, locomotor activity, feeding, body temperature, metabolism, immunological response, hormone production, and cognition, have a 24-h daily rhythm [2–4]. These rhythmic processes are governed by cell-autonomous circadian clocks found in nearly every cell of organisms. Circadian rhythm, however, may also be influenced by environmental cues and pathological situations, such as light, temperature, exercise, oxygen, food, eating patterns, cancer, cardiovascular disease, inflammation, and neurodegenerative illnesses [5–11]. The central pacemaker is located in the anterior hypothalamic suprachiasmatic nucleus (SCN) and regulates rhythms in peripheral clocks using a range of neuronal, humoral, and behavioral inputs [1]. Although peripheral clocks are autonomous, without the SCN, individual cells or tissues gradually desynchronize their cycles [12–14]. Although several cues have been suggested to aid in the communication between central and peripheral clocks, each tissue appears to respond to a different set of cues, most of which have not yet been fully explained [1]. Many of the cues involved in circadian clock communication are also found in immunological pathways (e.g., glucocorticoids (GCs) and cytokines). This shows that immunological responses may interfere with the control of the circadian clock [15].

Numerous cellular functions are regulated by the circadian rhythm, and these processes interact and influence biological activity. For instance, circadian clocks influence the cell cycle, immunological response, cognition, epigenetic modification, and redox homeostasis [15–17]. It is now well acknowledged that the inflammatory response, in particular the invasion and activation of immune cells, might play a substantial role in the development and maintenance of neuroinflammation. Circulatory myeloid and lymphoid cells, as well as tissue-resident cells, are examples of inflammatory cells. Microglia and astrocytes are tissue-resident inflammatory cells in the central nervous system (CNS) and are among the first to respond to injury/disease centrally [18]. These circulatory and CNS cells have been linked to a range of neuroinflammatory reactions, including multiple sclerosis and spinal cord damage. Cytokines and chemokines are mediators generated by these cells that can directly or indirectly activate or sensitize sensory neurons to induce or enhance the neuroimmune response [19]. While the literature on the role of inflammatory cells and their mediators in neuroinflammation outcomes has been extensively covered elsewhere, our focus here is to highlight the potential role of circadian rhythms in neuroimmune responses.

Immune function and neurodegeneration are linked to cellular metabolism [16, 17]. Disruptions of the Clock or its related proteins have been linked to a variety of clinical disorders that cause immune system malfunction, cancer, diabetes, obesity, and cognitive loss. Circadian disruption has also been linked to metabolic abnormalities, neurological illnesses, and microbial infections, all of which can lead to biological disorders [9]. Cellular circadian clocks have important functions in neuroinflammation and brain health, although these roles are currently poorly understood. Neuroinflammation and abnormal glial cell activation are common features of several neurodegenerative diseases, including Alzheimer's disease (AD) and Parkinson's disease (PD). Circadian disruption in sleep/wake cycles is one of the most prevalent and early symptoms of neurodegenerative disorders such as AD and PD, and it worsens the course of neurodegeneration [9]. Although the circadian clock has been linked to the control of peripheral inflammation, its significance in neuroinflammation is unclear. Here, we discuss the intricate circadian network and provide an overview of how the immune system and neuroinflammation are regulated by the circadian rhythm. The precise relationship between immune responses and clock mechanisms is still being studied, and progress in elucidating these mechanisms will be explored, with an emphasis on mammals.

2 Overview of the Mammalian Circadian System

In mammals, the circadian system is an internal timekeeping device that is hierarchical and structured in many oscillators at the organismal, cellular, and molecular levels [20, 21]. The central pacemaker is situated in the hypothalamic suprachiasmatic nucleus (SCN) and is made up of various populations of oscillating neurons and astrocytes that work together as a single circadian unit to produce a synchronized circadian signal [20–24]. Ablation of the SCN in mice disrupts rhythmicity in various processes, including behavior and endocrine function [20]. Further studies have shown that the body has other circadian oscillators, often known as "peripheral clocks," that are almost present in all cells of the body but that are dependent on the SCN for appropriate circadian rhythm entrainment and maintenance [25]. The SCN synchronizes peripheral clocks to a common 24-h rhythm and ensures that all systems in the body are synchronized and operating at peak performance. The SCN has two key characteristics that contribute to its "master pacemaker" potential. First, it is the only portion of the circadian system that is directly innervated by light-sensitive neurons. While the SCN is located deep within the brain, it receives external light cues via a retino-hypothalamic tract that passes via the optic nerve from the retina. This pathway plays a key role in the SCN's ability to be entrained by the light cycle, which works in conjunction with other signals to maintain the body's 24-h rhythm, such as temperature and food habits [26, 27]. Second, unlike peripheral clocks, connectivity between SCN neurons allows it to provide circadian outputs to practically every tissue and cell in the body [26]. In both the SCN and peripheral cells, these genes are involved in transcriptional-translational autoregulatory feedback loops that are cell-autonomous and highly conserved (Fig. 1).

The core molecular clock in each cell is made up of a positive transcriptional limb and a negative feedback limb. The bHLH-PAS transcription factor brain and muscle ARNT-like 1 (BMAL1) builds heterodimers with circadian locomotor output cycles Kaput (CLOCK) to promote circadian transcription by binding to E-box

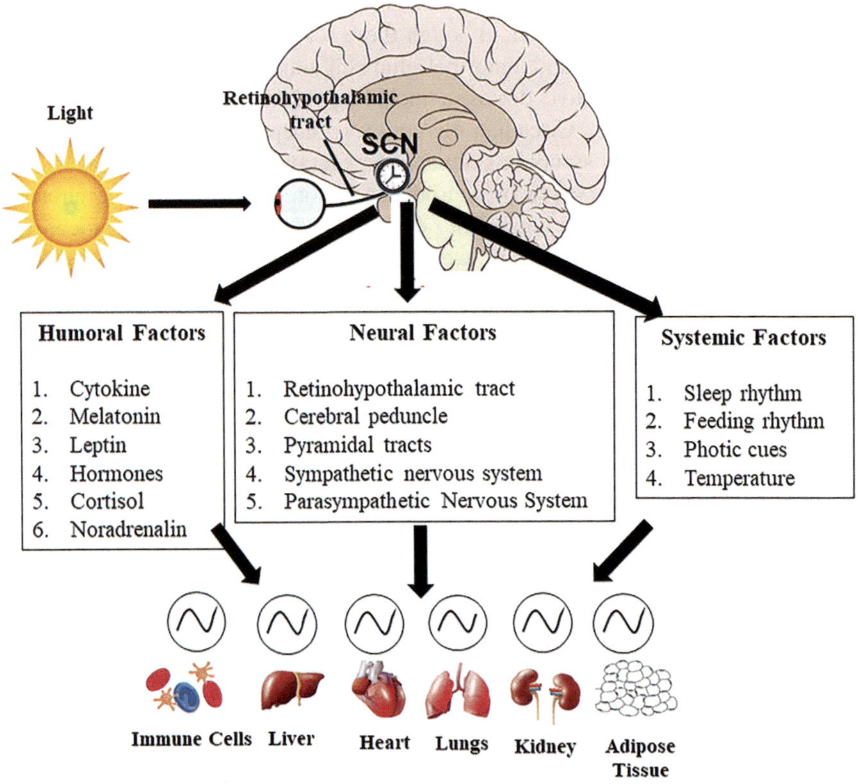

Fig. 1 Inflammation can impact the circadian system and its regulators. In mammals, light information captured by the retina is transmitted through the retino-hypothalamic tract to the SCN. Most peripheral tissues have their internal circadian clocks, which regulate their rhythms within the body. These rhythms are synchronized by the main SCN clock. The central clock regulates various factors such as humoral, neural, and systematic factors. This schematic diagram illustrates the communication between the central clock and peripheral clocks

motifs in the positive limb. The negative limb is made up of proteins from the Period circadian clock (Per1/2/3) and Cryptochrome (Cry1/2) families [2]. *Per* and *cry* proteins, in turn, inhibit the function of the BMAL/Clock complex, while the nuclear receptors *rora* and *rev-erba* regulate the transcription of *bmal1* and the nuclear factor interleukin-3 (*Nfil3*, also known as E4BP4) through opposing mechanisms of activation and repression, respectively [28]. This core clock is regulated to 24 h by a network of secondary clock proteins acting together to perform multiple post-translational procedures [29]. These core clock genes are expressed in almost every cell in the body and may produce circadian rhythms in transcription and cellular activity even in the absence of external signals. There are several transgenic mice models of circadian disruption, the most prevalent of which is global BMAL1 ablation or tissue-specific BMAL1 depletion utilizing the Cre-lox system. Similar research may be conducted by deleting additional circadian-associated genes such as *clock, npas2*, *per1*, and *per2*. However, because *bmal1* is a key component of the circadian positive regulatory loop, it is most efficient at modifying circadian rhythms when eliminated [30] (Fig. 2). Although studies have shown that CLOCK is a crucial component of the main loop, its homolog *npas2* may be able to make up for its absence; animals lacking *clocks* exhibit very modest circadian disruption [31]. The negative feedback loop of the circadian clock network requires a significant delay between transcription activation and repression to preserve daily oscillations of circadian clocks to about 24 h, which is achieved via post-translational modifications (PTMs). For example, the transcriptional repressor *per2* is phosphorylated by casein kinase Iε (CKIε) and targeted for ubiquitin-mediated degradation, whereas *cry1* is phosphorylated and destabilized by adenosine monophosphate-activated protein kinase (AMPK) [32, 33]. *Bmal1* transcription activator is phosphorylated by MAPK and glycogen synthase kinase-3β (GSK3β) [34], whereas Clock contains intrinsic histone acetyltransferase (HAT) activity, indicating a strong relationship between clock mechanism and chromatin remolding [35]. The stability and nuclear

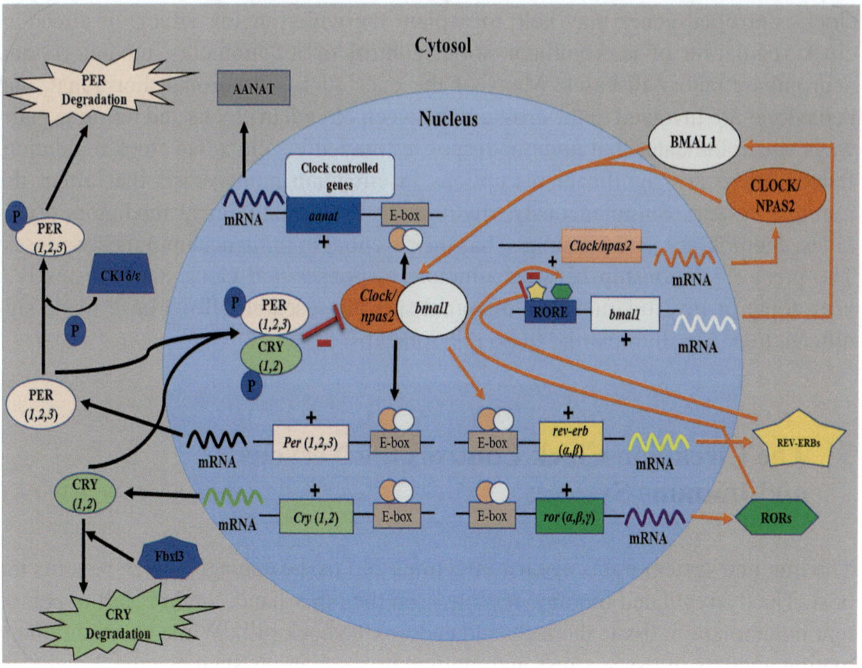

Fig. 2 The mammalian clock machinery. The autoregulatory feedback loops take 24 h to complete a cycle and constitute a circadian molecular clock. CLOCK and BMAL1 heterodimerize and initiate transcription including Period (*Per1*, *Per2*, and *Per3*) and Cryptochrome (*Cry1* and *Cry2*). Negative feedback is achieved by PER: CRY heterodimers that translocate back to the nucleus to repress their transcription by acting on the CLOCK: BMAL1 complex. REV-ERBα and RORα subsequently compete to bind ROREs present in the *Bmal1* promoter. Several clock-controlled genes (CCGs) and biological systems, including immunology and neuroinflammation function, are likewise rhythmically expressed by circadian clocks

translocation of core CLOCK proteins are influenced by PTMs controlled by phosphorylation, ubiquitylation, acetylation, and epigenetic modification, which in turn regulate the time-specific transcription of *clock* genes and play a vital role in the generation of circadian rhythm [36–38]. These findings suggest that non-transcriptional dependent redox oscillations play a critical role in the formation of circadian rhythms [39].

The circadian clock governs between 10 and 50% of all transcripts in a cell, depending on tissue type, and impacts essential processes such as cell cycle, redox homeostasis, inflammation, and metabolism [40]. Transcripts of inflammatory cytokines and host immunological responses vary during the day, and disturbance of the circadian clock leads to immune dysfunction [37, 41]. Furthermore, the circadian clock participates in a variety of metabolic processes in both peripheral tissues and the brain, including glucose transport, gluconeogenesis, lipogenesis, and mitochondrial processes such as morphological changes, respiration, oxidative phosphorylation, and mitochondrial biogenesis [40, 42–45]. As a result, disrupting the cell-autonomous clock network has a significant impact on immunological responses, energy consumption, and cognition in animals [43, 46]. This diversity of clock-controlled genes may help to explain the wide-ranging effects of circadian clock regulation of neuroinflammation, control of inflammatory mediators, and neuroinflammatory diseases. Many of the cues such as neuronal, hormonal, and behavioral are involved in the crosstalk between circadian clocks and immune pathways which indicates that immune responses may affect circadian clock regulation. Indeed, there are significant disruptions in circadian homeostasis that affect the immune system. Simultaneously, rhythmicity in immunological mediators (cytokines, chemokines, and hormones) has the potential to influence immune responses. The precise relationship between immune responses and clock mechanisms is a keen topic of investigation, and mammal-specific research efforts have made significant treads in illuminating these relationships.

3 The Circadian Clock Controls the Nervous and Immune Systems

The immune system fights against infections and tissue damage, which benefits the host. The dysregulated immune response, on the other hand, might result in persistent inflammation, tissue damage, and endotoxin shock [47]. While chronic inflammation appears in tissue stress and malfunction situations such as type 2 diabetes, obesity, and neurodegenerative and neuropsychiatric illnesses, the continuous low-grade inflammation contributes to the advancement of these diseases [48–50]. In mice, prolonged circadian phase changes (chronic jet lag) or simply changing the time of day [51, 52] can increase inflammation and LPS-induced endotoxemic mortality. Significant research has shown that the circadian clock regulates various immunological activities [26], including immune cell trafficking [53], pathogen identification, phagocytic capability [54], and the release of inflammatory cytokines, chemokines, and complement proteins. Immune mediators, on the other hand, can interact with *clock* genes and alter clock gene expression. Here we discuss the influence of circadian rhythms on neuroinflammation, especially innate and adaptive immunity, microglia, and myeloid and lymphoid immune cells [41, 54–56].

3.1 Adaptive and Innate Immunity

Clock genes such as *clock*, *per2* [57], *bmal1*, and the BMAL1 target *Nr1d1* all oscillate in peripheral macrophages [57, 58] and lymphocytes [59]. In humans, the blood levels of LPS-induced cytokines such as Interferon-γ (IFN-γ), Interleukin-10 (IL-10) [60], and TNF show a clear diurnal variation, whereas IL-6 levels fluctuate in an inconsistent manner [61, 62]. Mice exhibit time-of-day dependence for lymphocyte trafficking [59], LPS-induced monocyte recruitment, cytokine levels such as TNF, IL-6 [57], IL-12 [63], inducible nitric oxide synthase (iNOS—reactive NO-producing enzyme) [58], chemokines such as CCL5 [63], and mortality [64], with a decrease during late wake/early rest periods [57, 58]. *Bmal1* transcript displays a diurnal pattern in peritoneal macrophages, with a peak during the dark phase [41, 65, 66]. After being stimulated by bacterial endotoxin-lipopolysaccharide (LPS), peritoneal macrophages or bone marrow-derived macrophages (BMDMs) lacking *bmal1* produce more IL6, CXCL1, TNFα, and MCP1/CCL2 [53, 65]. Therefore, in a mouse model of diet-induced obesity, Bmal1 KO in monocytes decreases survival in response to infection and aggravates chronic inflammation and glucose intolerance [58]. *Bmal1* ablation also causes a Nrf2-dependent increase in ROS and IL-1 in macrophages [65]. Myeloid Bmal1 KO mice have higher blood IL-1b levels following LPS injection, as well as a higher mortality response to experimental or LPS-induced sepsis [41, 51, 65]. However, ablation of *bmal1* in macrophages protects mice from pneumococcal infection by increasing macrophage phagocytosis and decreasing pro-inflammatory cytokines in the lungs [67]. *Bmal1* is also implicated in chronic inflammation caused by monocytes and macrophages in diet-induced obesity and atherosclerosis [58, 68, 69]. Furthermore, *bmal1* controls the rhythmic expression of chemokine genes *Ccl2*, *Ccl8*, and *S100a8* in monocytes and peritoneal macrophages by binding to E-boxes in their promoters. Many studies have found that the clock has a role in regulating monocyte and macrophage activities. In mice, the phagocytic activity of macrophages was found to change over the day and night cycle [70]. Also, cytokine secretion after lipopolysaccharide (LPS) treatment of macrophages in vitro or LPS injection in mice has a diurnal rhythm, with stronger TNFα, IL-6, and other cytokine secretion in the early subjective night than in the early subjective day [57, 63]. Furthermore, Clock defective BMDMs display lower expression of pro-inflammatory genes *Il-6*, *Il1b*, *Tnfα*, *Ifn*-β, and *Ccl2* in response to LPS stimulation and decreased secretion of IL-6 and IL-1b in response to Salmonella infection [71]. *TNFα* and *IL-12* levels in peritoneal macrophages are reduced by the *per2* mutation in response to TLR9 ligand stimulation [72]. *Cry1* and *cry2* deficiency enhances *TNFα* and *IL*-6 production in BMDMs in response to LPS stimulation via binding to adenylyl cyclase and subsequently modulating cAMP levels, protein kinase A, and NF-kB activity [73]. Circadian rhythmicity in the abundance of the REV-ERBα transcription factor, which is regulated by the circadian clock in macrophages, has been shown to regulate a wide range of genes crucial for cytokine production and secretion [57, 63, 74]. These findings establish robust evidence for the circadian clock as an essential immunological regulator, which is a more complicated molecular process that may be underlying the circadian clock regulation of the innate immune response.

Additionally, the circadian clock influences adaptive immunity. *Rora* and *Rorγt* control pro-inflammatory T helper 17 (Th_{17}) cell development and increase the production of cytokine-interleukin-17 (IL-17) and IL-17F, whereas *rev-erba* inhibits *rorγt* activity by binding the same DNA motif [75–77]. Moreover, transcription repressor NFIL3 inhibits Th17 cell growth by directly binding to and inhibiting the *rorγt* promoter and is also required for the synthesis of IgE by activated B cells and the release of cytokines by Th1 and Th2 cells [78]. Furthermore, there is a diurnal fluctuation in lymphocyte migrating via lymph nodes and lymph, which is dependent on the rhythmic expression of promigratory factors on lymphocytes, and maybe loss of *bmal1* can be eliminated in lymphocytes [79, 80]. However, research found that the expression of *bmal1* did not affect T and B cell development or the adaptive immune response. Because of the extensive and complex effect of circadian clocks on peripheral immune response, it is vital to reconsider the role of the molecular clock at the systemic level, particularly in circadian disruption and immunological dysfunction-related neurodegenerative illnesses (Table 1).

Table 1 Innate and adaptive immunity mediated neuroinflammation

Type of interaction	Observed effects	References
Neuroinflammation mediated by innate immunity	1. Microglia are primed to become inflammatory after exposure to low light during the dark phase 2. Rat microglia's neuroinflammatory potential is modulated throughout the day, peaking in the light phase 3. Microglia with an enhanced inflammatory phenotype in *NR1d1* KO mice	[81, 82]
Neuroinflammation mediated by adaptive immunity	1. T-cell-specific ablation of *bmal1* reduces Experimental Autoimmune Encephalomyelitis (EAE) development after anti-myelin immunization throughout the day 2. The pro-inflammatory polarization of T cells is exacerbated by *bmal1* loss in myeloid cells 3. An increase in pro-inflammatory Th17 cell development and activity and a worsening of Experimental Autoimmune Encephalomyelitis are both caused by *Nr1d1* KO, which disinhibits *rorγt* 4. The pharmacological agonist *NR1D1/2* produces the opposite effects. *Rora* KO decreases Th17 response and lessens Experimental Autoimmune Encephalomyelitis; comparable results are seen following the administration of a *rora* pharmacological inhibitor	[19, 59, 83]

3.2 Lymphocytes

T and B lymphocytes, which are important effector cells of the adaptive immune system, have been widely researched in the context of pain response as well as biological rhythms [84–86]. Many aspects of their function as infection responders, such as migration to sites of inflammation [87], and cytokine production [88, 89], have been linked to circadian rhythms in some way. Numerous studies have discovered that the leukocyte populations in human blood have a circadian rhythm. The levels of CD41, or T helper cells, which secrete cytokines and stimulate other adaptive immune cells, and CD81, or cytotoxic T cells, which are activated to attack virus-infected or tumor cells, peak during the active phase, and melatonin injections can raise the CD41:CD81 T cell ratio [88–90]. Fortier et al. [89] discovered that lymph node-resident T cells collected at different times from mice housed in 24 h. darkness proliferate with a circadian rhythm following in vitro stimulation of the T cell receptor (TCR) with anti-CD3 antibody; both CD41 and CD81 T cells proliferated more during the subjective active (dark) phase.

Treatment with ionomycin and phorbol myristate acetate (PMA) at various time points showed that in vitro activation of $CD4^{+}$ T cells caused fluctuations in cytokine production and proliferation [88–90]. Other research has focused further upstream on T-cell activation pathways, that is, at the cellular level of the T-cell receptor and antigen delivery to T cells, even though PMA and ionomycin stimulate the proliferation of cells via intracellular signaling mechanisms (intracellular calcium, protein kinase C) [88, 89]. When T cells were triggered via their T-cell receptor with the mitogen concanavalin A [91] or an anti-CD3 T-cell receptor chain antibody, T-cell proliferation rhythms were observed [89]. The rhythmic expression of protein and mRNA by the tyrosine kinase ZAP70 downstream of the TCR is likely responsible for the rhythmic proliferation upon TCR activation. The researchers also examined the number of cells present in the lymph nodes without stimulation and discovered that neither B cells nor any subgroup of T cells changed in quantity during the day. T cells have clock genes; i.e. *e4bp4*, *per2*, *per3*, *rev-erbα*, and *rorα* exhibit circadian mRNA expression in human T cells isolated from blood, and rhythms for *bmal1*, *per3*, and *rev-erbα* were still found under in vitro condition [92]. Silver et al. [72] found rhythmic mRNA expression of *clock* genes in B cells from mouse spleen. *Per1* expression in mouse thymocytes, splenic T cells, and B cells is circadian, peaking at the start of the rest phase. *Per1* expression was lowest in thymocytes at the double-positive (DP) stage, but *bmal1* expression rose in the DP stage, most likely due to its negative regulation of *per1*. *Per1* expression varies throughout B cell development. This would imply that clock genes have a function in lymphocyte development. A B cell- or T cell-specific *bmal1* deletion, on the other hand, preserves normal B cell growth and function, as well as the amount of developing T cells [62].

Dendritic cells bearing antigens were used to immunize mice, and daytime antigen-specific activation was significantly higher than nighttime activation [89]. Subsequently, Yu et al. [93] discovered that the development of the T_H17 subtype depends on the 24-h clock in T cells. T_H17 cells help to protect mucosal surfaces from fungal and bacterial infection. In the intestine, T_H17 does not oscillate in frequency but it is exposed to an altered light cycle at higher frequencies; in REV-ERBα knockout mice this effect is obliterated. The number of and differentiation into T_H17 cells were found to be reduced in a CLOCK knockout, which inhibits the BMAL1-CLOCK transcriptional feedback loop. Naive T cells isolated from wild-type mice during the day were more likely to adopt the T_H17 profile after in vitro polarization than those isolated during the active phase (night). These findings might be explained by REV-ERBα inhibiting NFIL3, a TH17 repressor, influencing differentiation into the T_H17 profile [93]. These investigations demonstrate that T-cell responses to antigen presentation, subsequent proliferation of cells, development of regulatory function, and differentiation into several T-cell subtypes are all controlled daily. The remainder of this chapter will focus on the immune framework, the inflammatory response, and how it interacts with the circadian clock mechanism, even though these studies on rhythm in reaction to antigen presentation are significant and may help with better prevention and control of infectious diseases and more effective vaccination programs. Corresponding to this, other research conducted over the past 10 years has shown a complex relationship between the innate immune system's cells known as NK cells and macrophage, and the circadian clock.

3.3 Circadian Function on Microglia and Astrocytes

Glial cells, which comprise astrocytes, microglia, and oligodendrocytes, are the most numerous in the central nervous system (CNS). Glial cells are essential participants in CNS activity because they regulate critical pathways that alter the transmission of neuronal activity and also influence synapses, plasticity, axonal conductivity, and blood flow [94, 95]. Astrocytes, microglia, and oligodendrocytes appear to be the best cell types for regulating daily oscillation in physiological needs since glial cells also provide brain tissues with the metabolites required for life. Despite this, the possible role of glial cells in the formation and maintenance of circadian rhythms has only currently been recognized. One of the most important aspects of microglia is their capacity to dynamically sample the environment to manage tissue homeostasis. During CNS injury and disease, microglia act as a primary innate immune cell that produces and secretes several cytokines. Though extensive research has been shown to better understand the circadian rhythm of immune cells, limited studies have focused on microglia [96]. Many inflammatory signaling molecules that are linked to microglia activation have a circadian rhythm of expression. Circadian control of inflammatory genes has also been revealed in the spinal cord; however, the rhythm varies in the hypothalamus. Bmal1 KO mice showed reduced inflammatory cytokine expression from microglia compared to microglia isolated from WT mice [97]. Similarly, knocking down Bmal1 KO mice with siRNA transfection in microglial cell culture lowers IL-6 production as well as the expression of other inflammatory cytokines [19, 56] during an LPS-induced inflammatory response. In rat hippocampal microglia, the expression patterns of *per1*, *per2*, *rev-erbα*, and *bmal1* exhibit rhythmic fluctuations, particularly notable at the onset of the light phase [81]. In parallel, the *rev-erbα* agonist SR9011 reduces cytokine expression in primary cultures of rat brain microglia in response to immune activation, except the anti-inflammatory cytokine IL-10, which increases [98]. Taken together, enhanced *rev-erbα* signaling or *bmal1* deficiency lowers inflammatory cytokine release in a cell-autonomous manner. Microglia isolated from CLOCK mutants fail to maintain rhythmic expression of *clock* genes, which is crucial for regulating the circadian rhythm and the transcriptional-translational feedback loop [70]. TNFα, IL-6, and IL-1β are the most important pro-inflammatory cytokines released by microglia. TNF receptors are widely expressed, and their activation is critical in inducing neuronal death and apoptosis. IL-1β triggers several enzyme cascades, causing sphingomyelin hydrolysis, which can activate apoptotic pathways. IL-6 release stimulates the innate immune system, enhances cell survival, and increases cell division [99]. Cathepsin S is the important protease that is only expressed in the CNS microglia and plays a vital role in neuropathic pain and exhibits circadian rhythm in microglia and, in turn, regulates the rhythmic expression of clock genes in cortical microglia [70].

Astrocytes, the most numerous glial cells in the brain, play an important role in CNS activities, including those of the SCN. For example, whereas SCN neurons are extremely metabolically active during the light phase, SCN astrocytes are more active during the dark phase, inhibiting neuronal activity [100]. The reactive state of astrocytes can be linked to changes in their molecular, cellular, and functional properties. Astrogliosis can be useful or detrimental depending on the reactive state and clinical environment. Since the discovery of astrocyte clock gene expression in the cortex, spinal cord, and SCN, there has been increased interest in the link between astrocyte function and circadian rhythms [101, 102]. The astrocytes in older Bmal1 KO mice have thicker and longer branches, which are characteristics of astrogliosis, suggesting two different mechanisms are involved in the regulation of branches and encompassing the synapses [83]. The locomotor (wheel-running) activity can be determined by rhythms in the SCN astrocytes, and an astrocyte-specific BMAL1 deletion will increase the period of this locomotor rhythm [101]. Because *bmal1* deletion usually results in astrogliosis, one would anticipate Nr1d1 deletion to prevent astrogliosis because its equivalent proteins, REV-ERBα/β, block BMAL1 production. Astrogliosis was not found in Nr1d1 KO mouse astrocyte cultures or WT astrocyte cultures following Nr1d1 knockdown by siRNA [103]. A *per1/per2* double-mutant animal displayed broad behavioral arrhythmicity, but unlike age-matched Bmal1 KO or *Npas2/clock* KO mice, the double-mutant mice displayed no symptoms of astrogliosis [104]. Astrocytes also release essential immune mediators such as IL-6, IL-1β, and CCL2 in a circadian way, which induces a cascade of reactions leading to neuroinflammation and maintaining the strength of immune cells. Per1 KO increases the expression of IL-6 and CCL2 by cultured spinal astrocytes [105]. These in vitro findings showed that cytokine release by astrocytes, like morphology of astrocytes and cytoskeletal protein glial fibrillary acid protein (GFAP) expression, which is a major indicator of astrogliosis is a cell-autonomous process. The spinal astrocytes were controlling mechanical hypersensitivity by releasing ATP through glucocorticoids, which is mediated by upregulating the expression of serum- and glucocorticoid-inducible kinase 1 [106]. Many investigations have been conducted on the release of ATP by astrocytes, which exhibits a circadian oscillation that is controlled by mitochondrial calcium [52]. It has been reported that spinal astrocytes become activated following central nervous system (CNS) and peripheral nerve injuries. IL-1 enhances blood-brain barrier permeability in astrocytes by downregulating a key protein called sonic hedgehog [106]. Therefore, a disruption of the astrocytic clock system might have significant effects on the organism, such as neurodegeneration and neuroinflammatory diseases.

4 Circadian Control of Inflammatory Mediators

Cytokines (IL-2, IFN-γ, TNF-α), chemokines (CCL2 and CCL8), and hormones (cortisol, melatonin, prolactin, glucocorticoid, and norepinephrine) are the inflammatory mediators of immune cell activity. The expression pattern of these factors can be controlled by the CLOCK: BMAL1 positive transcriptional feedback mechanism [107, 108]. The activation of p65 by CLOCK, which in turn stimulates NF-κB, a protein complex responsible for regulating the release of immune components such as cytokines and chemokines, is a frequent mechanism of this gene regulation [64]. CRY may also control this process by inhibiting protein kinase A, which reduces p65 phosphorylation. Activation of NF-κB is reduced due to CLOCK inhibition by BMAL1, on the other side IκB which is an inhibitor of NF-κB, expression is increased by BMAL1 [53]. Furthermore, according to Nguyen [58], BMAL1 may bind E-boxes in the promoter regions of immunological mediator genes such TLR-9, CCL-2, and CCL8 as well as CLOCK can acetylate and show reduced binding capacity to the glucocorticoid receptor. This ligand-binding to promoters also exhibits a circadian rhythm, rising during the day and decreasing at night. A better knowledge of these interactions with neuroimmune disease may aid in the direction of therapy and therapeutic choices.

Pro-inflammatory cytokines such as IL-2, IFNγ, and TNFα, as well as the anti-inflammatory cytokine IL-10, are secreted at the highest levels during the resting phase. Meanwhile, IL-6 levels in plasma follow a biphasic (12-h) cycle, whereas levels in cerebrospinal fluid follow a daily rhythm [109]. Chemokines such as CCL2 and CCL8 are also reported to have a diurnal production when attracting monocytes to areas of inflammation [80], although other chemokines such as CCL3, CCL5, CXCL1, CXCL2, CX3CL1 were not showing this activity [3]. CCL2 is expressed at a maximum level in the SCN during the active phase. It is unclear if the time of day of activation affects the cytokine response. It has been reported that different time schedules of LPS administration elicited varying amounts of IL-1α but not IL-1β, TNF-α, or IL-6 as well as the chemokines CCL5, CCL12, and CXCL1 were all generated at variable quantities due to different schedule [63]. As a result, many chemokines and cytokines exhibit diurnal variation in disease states or when immune cells are activated, and their control is frequently model-dependent.

The impact of *clock* gene knockouts, which result in the lack of a functioning circadian system, on the production of cytokines has been the subject of several research. IFN-γ, perforin, and granzyme B rhythms in isolated NK cells are changed in *per1/2* mice, and a polymorphism in the *per3* gene is related to higher levels of IL-6 [110]. Following LPS stimulation, bone marrow transplantation from a CRY double-knockout mutant mouse showed maximum iNOS, TNF-α, and IL-6, which was likely mediated via the NF-κB pathway, showing the involvement of clock genes in immune system regulation [73]. According to Nguyen [58], BMAL KO mice increased the gene expression of CCL2 and CCL8 in monocytes and peritoneal macrophages, which resulted in greater serum protein levels, and it interfered with the rhythmic expression of CXCL5. Furthermore, REV-ERBα overexpression reduces CCL2 induction after LPS exposure, and macrophages isolated from REV-ERBα knockout animals have enhanced CCL2 expression [55]. IL-2 production remained rhythmic in a T cell-specific BMAL knockout mice model, indicating that cytokine rhythms are mediated by cell-extrinsic rather than cell-intrinsic circadian signals. In overview, many cytokines and chemokines are produced with a circadian rhythm, and the timing of an immune response or the lack of clock genes can dramatically impact the amplitude of a cytokine response.

Cortisol and noradrenaline, two rhythmic hormones, appear to play a part in regulating the rhythm of immune cell population abundance. Hormones, notably glucocorticoids, play an important role in the circadian entrainment of peripheral clocks: dexamethasone administration increases circadian gene expression in cultured rat fibroblasts but not SCN neurons [111]. According to [144], [145], cortisol injections in humans cause a decrease in the blood levels of $CD4^+$ and $CD8^+$ impulsive, central memory, and effector memory T cells. However, when cortisol levels are pharmacologically lowered or a GC receptor antagonist is employed, the opposite effect is shown. The expression of the chemokine receptor CXCR4 in these cells, which mediates T-cell redistribution (bone marrow homing) in response to cortisol, is negatively linked with these cortisol effects on T-cell population rhythms [112]. Cortisol levels in human plasma oscillate between a low during the rest phase (nighttime) and a maximum at the time of morning. ACTH and cortisol serum concentrations are twice in the morning compared to the evening but mRNA expression on glucocorticoid receptors is greater in the evening than in the morning [113]. The high expression of the chemokine receptor CX3CR1 in $CD8^+$ effector T cells suggests that noradrenaline facilitates their demargination from the vascular endothelium. It has been observed that the daily rhythmicity of leukocytes showed in mice's skeletal muscle and bone marrow. The sympathetic nervous system (SNS) causes a daily oscillation of circulating chemokines, which is regulated by the central clock [3].

Circadian clocks have also been demonstrated to be influenced by glucocorticoids (GCs). GC receptor binding elements were discovered in the promoters of the clock genes *per1*, *per2*, and *rev-erbα* [114, 115]. GC rhythms are required for PER2 protein rhythms in the limbic forebrain. In terms of the SCN master clock, the absence or low amounts of GC receptors in the adult SCN most likely explains why GC injection or adrenalectomy has no influence on *clock* gene expression in the SCN [111, 116, 117]. The peak expression of PER1 is also delayed in various organs in rats after adrenalectomy, which lowers the hormone content in peripheral blood. Surprisingly, when exposed to disrupted light cycles, these rats regained their original *per1* phase faster than control animals. In general, GCs have been determined to be plausible candidates for mediating peripheral clock resetting, and they have been found to influence behavioral resetting, which is believed to be regulated by the SCN [118]. Considering the circadian regularity of GC levels, the suppression of NF-κβ is also likely to have a circadian rhythm. Despite the known impact of NF-κβ on clock genes, it is plausible that GCs impose a temporal dependency on the immunological control of circadian clock gene expression through regulating NF-κβ activity. These findings imply that variable GC formation in response to an immunological assault may impose time- and tissue-dependent control of clock gene expression.

Melatonin is a hormone that is produced from serotonin by the enzyme arylalkylamine N-acetyltransferase (AANAT) is the rate-limiting step and it is synthesized and secreted by the pineal gland [119]. According to Carillo-Vico et al. (2005), melatonin has immunomodulatory effects, and depending on the cell type and circumstances, it can play both pro- and anti-inflammatory roles in the body's inflammatory response. Melatonin supplementation has been found to reduce inflammation in multiple inflammatory models by reducing inducible nitric oxide synthase, PGE_2 levels, and pro-inflammatory cytokine levels [120]. Melatonin has been found to prevent LPS-induced NF-κβ activation in a microglial cell line, limiting chemokine release and increasing this hormone's anti-inflammatory function. During inflammation, NF-κβ appears to be a major factor in both the downregulation of AANAT in the pineal and its stimulation in immune cells, such as macrophages, resulting in melatonin release by these cells [121]. Both melatonin and corticosterone affect NF-κβ in the pineal gland, where NF-κβ protein levels are rhythmic and melatonin inhibits NF-κβ activation. Stress-induced plasma corticosterone also lowers NF-κβ nuclear levels in the hamster pineal gland [122, 123]. In general, the interaction between melatonin and GC cycles, which regulate inflammation, may be a factor in the pro-inflammatory states brought on by an acute inflammatory stimulus.

5 Molecular Circadian Clock Intervention in Neurodegeneration and Inflammation

Neurodegeneration is primarily defined by the gradual loss of a neuron's structure or function, which ultimately results in neuronal death. Recent research reveals that many cell types (astrocytes and microglia) in the brain have a role in neurodegenerative progression [124]. Therefore, as shown by several recent research, altering the brain's astrocytes' and microglia's internal clocks may be a unique therapeutic approach for treating neurodegenerative disorders. Astrocytes control the transmission of neuronal signals, whereas microglia offer immune surveillance and alter neural circuitry. Chronic neuroinflammation caused by activated microglia and astrocytes is important in the genesis and progression of neurodegeneration [125]. The function of brain cells such as neurons, astrocytes, microglia, and endothelial cells is directly impacted by the circadian clock, according to compelling research [126–128]. Circadian clock disruption is a major risk factor for the development of multiple neurodegenerative diseases, and individuals with neurodegenerative illness frequently have disrupted circadian rhythmicity. The most prevalent age-related neurodegenerative illnesses are AD and PD, which have significant circadian disruption [129]. The clock-controlled gene *Chi3l1*, which is produced mostly by brain astrocytes, encodes a glycoprotein that is overexpressed in Alzheimer's disease patients and a double transgenic mouse model of the illness [130]. The ablation of the gene *Chi3l1* inhibits the buildup of amyloid plaques, which is the primary symptom of Alzheimer's disease, and global deletion of *bmal1* reduces *Chi3l1* expression in healthy mice [130]. The deposition of amyloid plaques, which are made up of aggregated amyloid-beta (Aβ) peptides, is considered to be the first step in the pathogenic processes that cause AD. This is followed by the development of tau protein aggregates inside neurons [131]. In astrocytes, the upregulated *bmal1* expression suggests the cause of elevated *Chi3l1* expression that results in amyloid plaque buildup. In a transgenic mouse model of AD, deletion of *bmal1* induces severe circadian fragmentation, dramatically blunts Aβ rhythms, and promotes amyloid plaque formation [132]. This theory is supported by a human investigation, which discovered that the levels of the proteins BMAL1 and CLOCK were abnormally high in the cerebral cortical astrocytes of individuals with Alzheimer's disease [14]. The interruption of clock gene expression in Alzheimer's disease is not limited to astrocyte cell type. The circadian expression levels of clock genes and pro-inflammatory genes in cortical microglia were irregular in the amyloid precursor protein knock-in (APP-KI) mouse model showing an early stage of the illness [133]. The gene expression is increased in the hippocampus by global *bmal1* and *rev-erbα* knockout, and in the Bmal1 KO group increased C4b transcript levels in the neurons and astrocytes of the cerebral cortex [82, 103]. Inhibiting *rev-erbα*, on the other hand, decreases amyloid plaque load and protects synaptic loss in 5XFAD mice, an animal model of AD that may be caused by enhanced microglial Aβ clearance [66, 82]. Targeting *bmal1* expression in glial cells could mitigate or perhaps prevent illness. Previously published data supported the hypothesis that disturbance of the clock machinery in microglia and astrocytes drives neuropathogenesis (Table 2).

Another neurodegenerative disorder could involve the microglial circadian clock system. Global Bmal1 KO accelerates the death of dopaminergic neurons in the substantia nigra, exacerbates dopamine depletion, and worsens motor deficits in a mouse model of Parkinson's disease [141]. Further research indicated that substances generated by cultured microglia with *bmal1* deletion accelerated the death of midbrain dopaminergic neurons under in vitro conditions [141]. The implications of *bmal1* depletion in astrocytes may also be complicated, and worthwhile in the

Table 2 Various studies have associated AD, PD, and neuroinflammation factors with the molecular clock mechanism

Risk factors	Clock Gene (s)	Observed effects	References
Alzheimer's disease (AD)	*Bmal1, Per2*	1. Irregular oscillations in the SCN of 5XFAD animals were strongly influenced by *bmal1* and *per2* mRNA levels 2. In vitro, AD causes the degradation of *bmal1* and CBP via SUMOylation 3. Disruptions of the oscillation of *per2* expression levels 4. SUMO1 knockdown salvaged BMAL1 degradation in cells	[134]
	bmal1, per2	1. While musculoskeletal *BMAL1* levels are unaffected by aging, SCN BMAL1 expression is reduced 2. Modified hippocampal *bmal1* expression 3. Altered hippocampal and hypothalamic *per2* expression	[135, 136]
Aging, neuroinflammation	*bmal1, clock, npas2, per1, per2*	1. *bmal1*, *clock*, and *npas2* gene ablation resulted in severe reactive astrogliosis 2. Synaptic terminal degeneration, neuronal oxidative damage, and reduced expression of many redox defense genes were all caused by the deletion of BMAL1	[104]
	bmal1, clock, cry1, cry2, per1, per2	1. The impairment of *per* and *cry* expression by BMAL1 KO results in an increase in ROS generation and persistent oxidative stress in the brain 2. A lack of *bmal1*, *clock*, *cry1*, or *cry2* may affect open-field behavior, exploratory activity, or habit	[137]

Parkinson's disease (PD)	*bmal1, clock, cry1, per2, rora*	1. In 6-OHDA rats, *rora* mRNA levels increased while *bmal1*, *per2*, and *clock* expression levels decreased 2. In 6-OHDA-treated cells, *bmal1*, *clock*, *per2*, and *rora* expression levels were reduced, BMAL1/CLOCK protein expression was reduced, and the BMAL/CLOCK binding ratio was increased 3. Resveratrol partially restores acetylated *bmal1* levels, *cry1*, and *per2* expression levels in cells	[29]
	bmal1, clock, cry1, cry2, per1, per2, per3	1. *clock*, *cry1*, and *cry2* levels were elevated and PER2 expression was downregulated in P1 fibroblasts 2. P2 fibroblasts reduce *per3* and *cry2* expression 3. Significant oscillatory amplitude alterations were seen in *cry1*, *per2*, and *per3* for both cases	[138]
	bmal1, clock, cry1-2, npas2, nr1d1, rorβ	1. The risk of PD is highly correlated with *bmal1* and *per1* 2. *bmal1* and *per1* SNPs are also linked to PD	[139]
PD, neuroinflammation	*bmal1, clock, cry1, cry2, per1, per2, rev-erbα, rorα*	1. Dopaminergic neuron loss and significant neuroinflammation occurred in LPS-injected rats 2. *clock* and *npas2* expression levels decreased in all groups, whereas *bmal1* protein and mRNA expression decreased in all groups 3. While *cry1* and *cry2* expressions were unaffected in any group, *per1* and *per2* dropped in all of them 4. In all groups, the expression of the proteins *rev-erbα* and *rorα* reduced, as did their gene expression levels	[140]

case of neurodegenerative illness, but not in other cases. It is well known that the expression of genes involved in the absorption of fatty acids and glucose by microglia varies depending on the time of day; for example, the expression of the genes *Lpl* and glucose transporter member 5 (Glut5) is higher during the dark phase, when microglia are more active [97]. *Glu5* and *Lpl* expression, as well as *Il1b*, are all reduced in microglia when *bmal1* is knocked down [97]. Thus, immune oscillations may be triggered by periodic variations in cellular metabolism. As previously stated, *bmal1* regulates cytokine release in both glial cells independently. Because its deletion reduced the release of pro-inflammatory chemicals in microglia and astrocytes, *bmal1* depletion in mice should have the same effect [142]. We thus propose that the organism is more significantly impacted by the structural alterations in astrocytes brought about by Bmal1 KO than by the decrease in the production of pro-inflammatory cytokines. Thus, maintaining the molecular clock network's and cellular metabolism's homeostasis is critical for optimal immune function and the prevention of neurodegeneration.

6 Translational Perspectives

Circadian rhythms and neuroinflammation interact at the molecular, cellular, and pathological levels, although this area of research is still relatively unexplored. The correlation between circadian rhythms and neuroinflammation has been demonstrated by genetically engineered studies. However, the underlying processes of these phenomena remain poorly understood, despite some recent evidence offering some clues. Genetic research employing large patient populations, such as the UK Biobank, can help find novel possible relationships between genes affecting circadian rhythms and pain, or vice versa. Alternatively, changes in pain manifestations have been identified between mouse strains; whether these are attributable to differences in circadian rhythms and neuroimmune mediator expression is uncertain. Recent research utilizing floxed transgenic mice has demonstrated that genes like BMAL1 regulate the activation of immune cells and the generation of mediators. However, further research is needed to determine the direct or indirect role of circadian-controlled genes in neuroinflammation. Cytokines, chemokines, Leptin, PGE2, GCs, and melatonin are examples of inflammatory mediators that may help to control the rhythmicity of the circadian clock as well as immune response. Furthermore, in inflammatory situations, tissue-specific alterations to circadian clocks may be imposed through direct neural connections from the brain to the periphery. To maintain brain homeostasis, circadian rhythm integrity, which includes sleep and metabolic cycle competence, must be maintained. However, this integrity deteriorates with age and during neurodegeneration. Future studies should concentrate on the increased temporal and contiguous distinction of circadian rhythms on brain glial cells, which may be achieved by using cutting-edge methods. Global gene-deletion approaches have until now been the most common strategy employed, although they can introduce certain confusing factors throughout expansion and in glial cells at various CNS regions. CRISPR-Cas9, a technique that is constantly evolving, makes it feasible to selectively disrupt genes in glial cells at particular loci [143]. More research targeting clock genes in other areas of the CNS is likely in the future years. Numerous studies have shown that the primary pathway connecting circadian disruption and neurodegenerative illness is stress and neuroinflammation. The correlation between CK1d and pathogenic tau deposits as well as the functions of *Rev-erbα* in the neuroinflammatory response are two of the findings that may be the most intriguing. *Rev-erbα* plays a key role in regulating microglial activation and neuroinflammation, positioning it as a promising pharmaceutical target for neuroinflammatory and neurodegenerative diseases. Advancing therapies that modulate *rev-erbα* could represent a significant breakthrough in slowing the progression of neurodegenerative disease and managing neuroinflammation. A therapeutic approach in diseased circumstances can involve focusing on the clock machinery in various cell types. Although genetically engineered mouse lines are an effective technique for increasing our understanding of biological systems, utilizing drugs in preclinical investigations enables a more direct application to human research.

7 Conclusion

The intricate interplay between the circadian clock system and immune function, particularly within glial cells, represents a critical frontier in neuroscience research with substantial therapeutic implications. Current evidence underscores the pivotal role of glial cell circadian rhythms in maintaining immunological homeostasis and modulating the activity of neurons, astrocytes, and endothelial cells. Disruptions in these rhythms are associated with aberrant glial cell morphology and altered neuroimmune responses during inflammation and tissue damage. Notably, astrocytes within the suprachiasmatic nucleus (SCN) have been shown to independently influence circadian behaviors, and targeted knockout of the *bmal1* gene in glial cells further elucidates their impact on circadian regulation.

The circadian modulation of immune and cognitive functions appears to be mediated through intricate cellular metabolic pathways. The recognition of glial cells as integral components of the circadian system highlights their potential as targets for therapeutic intervention in various neurological conditions. A more comprehensive understanding of glial circadian rhythms and their functional significance is crucial for developing effective treatments for brain injuries and neurodegenerative diseases. This understanding could enable earlier detection and intervention in neurodegenerative diseases by clarifying the mechanistic links between the circadian clock machinery, inflammation, and neurodegeneration. Thus, advancing research in this rapidly expanding domain is essential for unlocking new therapeutic avenues and improving patient outcomes in neurodegenerative and neuroinflammatory disorders.

Compliance with Ethical Standards

Conflict of Interest The authors declare that they have no known competing financial interests or personal relationships that could have appeared to influence the work reported in this paper.

References

1. Dibner C, Schibler U, Albrecht U. The mammalian circadian timing system: organization and coordination of central and peripheral clocks. Annu Rev Physiol [Internet]. 2010;72(1):517–49. Available from: https://doi.org/10.1146/annurev-physiol-021909-135821
2. Mohawk JA, Green CB, Takahashi JS. Central and peripheral clocks. Annu Rev Neurosci. 2012;35:445–62.
3. Scheiermann C. Nihms-599947. Nat Rev Immunol. 2014;13(3):190–8.
4. Gnocchi D, Bruscalupi G. Circadian rhythms and hormonal homeostasis: Pathophysiological implications. Biology (Basel). 2017;6(1):1–20.
5. Cavadini G, Petrzilka S, Kohler P, Jud C, Tobler I, Birchler T, et al. TNFa suppresses the expression of clock genes 2007 PNAS.pdf. Proc Natl Acad Sci U S A. 2007;104(31):12843–8.
6. Pivovarova O, Jürchott K, Rudovich N, Hornemann S, Ye L, Möckel S, et al. Changes in dietary fat and carbohydrate content alter central and peripheral clock in humans. J Clin Endocrinol Metab. 2015;100(6):2291–302.
7. Adamovich Y, Ladeuix B, Golik M, Koeners MP, Asher G. Rhythmic oxygen levels reset circadian clocks through HIF1α. Cell Metab [Internet]. 2017;25(1):93–101. Available from: https://doi.org/10.1016/j.cmet.2016.09.014
8. Hood S, Amir S, Hood S, Amir S. The aging clock : circadian rhythms and later life Find the latest version : The aging clock : circadian rhythms and later life. J Clin Invest. 2017;127(2):437–46.
9. Hood S, Amir S. Neurodegeneration and the circadian clock. Front Aging Neurosci. 2017;9(MAY):1–9.
10. Crosby P, Hamnett R, Putker M, Hoyle NP, Reed M, Karam CJ, et al. Insulin/IGF-1 drives PERIOD synthesis to entrain circadian rhythms with feeding time. Cell [Internet]. 2019;177(4):896–909.e20. Available from: https://doi.org/10.1016/j.cell.2019.02.017
11. Gabriel BM, Zierath JR. Circadian rhythms and exercise — re-setting the clock in metabolic disease. Nat Rev Endocrinol [Internet]. 2019;15(4):197–206. Available from: https://doi.org/10.1038/s41574-018-0150-x
12. Nagoshi E, Saini C, Bauer C, Laroche T, Naef F, Schibler U. Circadian gene expression in individual fibroblasts: Cell-autonomous and self-sustained oscillators pass time to daughter cells. Cell. 2004;119(5):693–705.
13. Yamazaki S, Numano R, Abe M, Hida A, Takahashi R, Ueda M, et al. Resetting central and peripheral circadian oscillators in transgenic rats. Science [Internet]. 2000;288(5466):682–5. Available from: https://doi.org/10.1126/science.288.5466.682
14. Yoo SH, Yamazaki S, Lowrey PL, Shimomura K, Ko CH, Buhr ED, et al. PERIOD2::LUCIFERASE real-time reporting of circadian dynamics reveals persistent circadian oscillations in mouse peripheral tissues. Proc Natl Acad Sci U S A. 2004;101(15):5339–46.
15. Cermakian N, Westfall S. Circadian clocks and inflammation: reciprocal regulation and shared mediators. Arch Immunol Ther Exp (Warsz). 2014;62:303–18.
16. Vijayan V, Pradhan P, Braud L, Fuchs HR, Gueler F, Motterlini R, et al. Human and murine macrophages exhibit differential metabolic responses to lipopolysaccharide—a divergent role for glycolysis. Redox Biol [Internet]. 2019;22(January):101147. Available from: https://doi.org/10.1016/j.redox.2019.101147

17. Wang L, Pavlou S, Du X, Bhuckory M, Xu H, Chen M. Glucose transporter 1 critically controls microglial activation through facilitating glycolysis. Mol Neurodegener. 2019;14(1):1–15.
18. Segal JP, Tresidder KA, Bhatt C, Gilron I, Ghasemlou N, Brécier A, et al. Circadian control of pain and neuroinflammation. J Neurosci Res [Internet]. 2021;96(6):1002–20. Available from: https://doi.org/10.1002/jnr.24150
19. Wang XL, Li L. Circadian clock regulates inflammation and the development of neurodegeneration. Front Cell Infect Microbiol. 2021;11(September):1–16.
20. Reppert SM, Weaver DR. Coordination of circadian clocks in mammals. Nature [Internet]. 2002;418(8):935–41. Available from: https://www.nature.com/articles/nature00965
21. Saini R, Jaskolski M, Davis SJ. Circadian oscillator proteins across the kingdoms of life: Structural aspects. BMC Biol. 2019;17(1):1–39.
22. Hastings MH, Maywood ES, Brancaccio M. Generation of circadian rhythms in the suprachiasmatic nucleus. Nat Rev Neurosci [Internet]. 2018;19(8):453–69. Available from: https://doi.org/10.1038/s41583-018-0026-z
23. Herzog ED, Hermanstyne T, Smyllie NJ, Hastings MH. Regulating the suprachiasmatic nucleus (SCN) circadian clockwork: Interplay between cell- autonomous and circuit-level mechanisms. Cold Spring Harb Perspect Biol. 2017;9(1):1–26.
24. Brancaccio M, Edwards MD, Patton AP, Smyllie NJ, Chesham JE, Maywood ES, et al. Cell-autonomous clock of astrocytes drives circadian behavior in mammals. Science. 2019;363(6423):187–92.
25. Duguay D, Cermakian N. The crosstalk between physiology and circadian clock proteins. Chronobiol Int. 2009;26(8):1479–513.
26. Man K, Loudon A, Chawla A. Immunity around the clock. Science. 2016;354(6315):999–1003.
27. Yoshida K, Hashimoto T, Sakai Y, Hashiramoto A. Involvement of the circadian rhythm and inflammatory cytokines in the pathogenesis of rheumatoid arthritis. J Immunol Res. 2014; https://doi.org/10.1155/2014/282495.
28. Duez H, Staels B. The nuclear receptors Rev-erbs and RORs integrate circadian rhythms and metabolism. Diabetes Vasc Dis Res. 2008;5(2):82–8.
29. Wang J, Symul L, Yeung J, Gobet C, Sobel J, Lück S, et al. Circadian clock-dependent and -independent posttranscriptional regulation underlies temporal mRNA accumulation in mouse liver. Proc Natl Acad Sci U S A. 2018;115(8):E1916–25.
30. Bunger MK, Wilsbacher LD, Moran SM, Clendenin C, Laurel A, Hogenesch JB, et al. Mop3 is an essential component of the master circadian pacemaker in mammals. Cell. 2013;103(7):1009–17. Available from: https://www.ncbi.nlm.nih.gov/pmc/articles/PMC3779439/pdf/nihms517125.pdf
31. Dardente H. Synchronization and genetic redundancy in circadian clocks. Medicine/Sciences. 2008;24(3):270–6.
32. Eide EJ, Woolf MF, Kang H, Woolf P, Hurst W, Camacho F, et al. Control of mammalian circadian rhythm by CKIε-regulated proteasome-mediated PER2 degradation. Mol Cell Biol. 2005;25(7):2795–807.
33. Lamia KA, Sachdeva UM, Ditacchio L, Williams EC, Alvarez JG, Egan DF, et al. AMPK regulates the circadian clock by cryptochrome phosphorylation and degradation. Science. 2010;326(5951):437–40.
34. Sahar S, Zocchi L, Kinoshita C, Borrelli E, Sassone-Corsi P. Regulation of BMAL1 protein stability and circadian function by GSK3β-mediated phosphorylation. PLoS One. 2010;5(1)
35. Doi M, Hirayama J, Sassone-Corsi P. Circadian regulator CLOCK is a histone acetyltransferase. Cell. 2006;125(3):497–508.
36. Gallego M, Virshup DM. Post-translational modifications regulate the ticking of the circadian clock. Nat Rev Mol Cell Biol. 2007;8(2):139–48.
37. Bellet MM, Sassone-Corsi P. Mammalian circadian clock and metabolism—the epigenetic link. J Cell Sci [Internet]. 2010;123(22):3837–48. Available from: https://doi.org/10.1242/jcs.051649

38. Shi G, Xie P, Qu Z, Zhang Z, Dong Z, An Y, et al. Distinct roles of HDAC3 in the core circadian negative feedback loop are critical for clock function. Cell Rep [Internet]. 2016;14(4):823–34. Available from: https://doi.org/10.1016/j.celrep.2015.12.076
39. Rey G, Reddy AB. Interplay between cellular redox oscillations and circadian clocks. Diabetes Obes Metab. 2015;17(S1):55–64.
40. Zhang R, Lahens NF, Ballance HI, Hughes ME, Hogenesch JB. A circadian gene expression atlas in mammals: Implications for biology and medicine. Proc Natl Acad Sci U S A. 2014;111(45):16219–24.
41. Deng W, Zhu S, Zeng L, Liu J, Kang R, Yang M, et al. The circadian clock controls immune checkpoint pathway in sepsis. Cell Rep. 2018;24(2):366–78.
42. Kawai M, Green CB, Lecka-Czernik B, Douris N, Gilbert MR, Kojima S, et al. A circadian-regulated gene, Nocturnin, promotes adipogenesis by stimulating PPAR-γ nuclear translocation. Proc Natl Acad Sci U S A. 2010;107(23):10508–13.
43. Schiaffino S, Blaauw B, Dyar KA. The functional significance of the skeletal muscle clock: Lessons from Bmal1 knockout models. Skelet Muscle [Internet]. 2016;6(1):1–9. Available from: https://doi.org/10.1186/s13395-016-0107-5
44. Robles MS, Humphrey SJ, Mann M. Phosphorylation is a central mechanism for circadian control of metabolism and physiology. Cell Metab [Internet]. 2017;25(1):118–27. Available from: https://doi.org/10.1016/j.cmet.2016.10.004
45. De Goede P, Wefers J, Brombacher EC, Schrauwen P, Kalsbeek A. Circadian rhythms in mitochondrial respiration. J Mol Endocrinol. 2018;60(3):R115–30.
46. Stenvers DJ, Scheer FAJL, Schrauwen P, la Fleur SE, Kalsbeek A. Circadian clocks and insulin resistance. Nat Rev Endocrinol [Internet]. 2019;15(2):75–89. Available from: https://doi.org/10.1038/s41574-018-0122-1
47. Medzhitov R. Origin and physiological roles of inflammation. Nature. 2008;454(7203):428–35.
48. Marogianni C, Sokratous M, Dardiotis E, Hadjigeorgiou GM, Bogdanos D, Xiromerisiou G. Neurodegeneration and inflammation—an interesting interplay in Parkinson's disease. Int J Mol Sci. 2020;21(22):1–15.
49. Mishra SP, Wang B, Jain S, Ding J, Rejeski J, Furdui CM, et al. A mechanism by which gut microbiota elevates permeability and inflammation in obese/diabetic mice and human gut. Gut. 2023;1–18
50. Fenneman AC, Weidner M, Chen LA, Nieuwdorp M, Blaser MJ. Antibiotics in the pathogenesis of diabetes and inflammatory diseases of the gastrointestinal tract. Nat Rev Gastroenterol Hepatol. 2023;20(2):81–100.
51. Curtis AM, Fagundes CT, Yang G, Palsson-McDermott EM, Wochal P, McGettrick AF, et al. Circadian control of innate immunity in macrophages by miR-155 targeting Bmal1. Proc Natl Acad Sci U S A. 2015;112(23):7231–6.
52. Marpegan L, Leone MJ, Katz ME, Sobrero PM, Bekinstein TA, Golombek DA. Diurnal variation in endotoxin-induced mortality in mice: correlation with proinflammatory factors. Chronobiol Int [Internet]. 2009;26(7):1430–42. Available from: https://doi.org/10.3109/07420520903408358
53. Curtis AM, Bellet MM, Sassone-Corsi P, O'Neill LAJ. Circadian clock proteins and immunity. Immunity [Internet]. 2014;40(2):178–86. Available from: https://doi.org/10.1016/j.immuni.2014.02.002
54. Timmons GA, O'Siorain JR, Kennedy OD, Curtis AM, Early JO. Innate rhythms: clocks at the center of monocyte and macrophage function. Front Immunol. 2020;11(August):1743.
55. Sato TK, Panda S, Miraglia LJ, Reyes TM, Rudic RD, McNamara P, et al. A functional genomics strategy reveals rora as a component of the mammalian circadian clock. Neuron. 2004;43(4):527–37.
56. Nakazato R, Kawabe K, Yamada D, Ikeno S, Mieda M, Shimba S, et al. Disruption of Bmal1 impairs blood-brain barrier integrity via pericyte dysfunction. J Neurosci. 2017;37(42):10052–62.

57. Keller M, Mazuch J, Abraham U, Eom GD, Herzog ED, Volk HD, et al. A circadian clock in macrophages controls inflammatory immune responses. Proc Natl Acad Sci U S A. 2009;106(50):21407–12.
58. Nguyen KD. Circadian gene Bmal1 regulates diurnal oscillations of Ly6C. Science. 2013;341(September):1483–8.
59. Druzd D, Matveeva O, Ince L, Harrison U, He W, Schmal C, et al. Lymphocyte circadian clocks control lymph node trafficking and adaptive immune responses. Immunity [Internet]. 2017;46(1):120–32. Available from: https://doi.org/10.1016/j.immuni.2016.12.011
60. Petrovsky N, Harrison LC. Diurnal rhythmicity of human cytokine production. J Immunol. 1997;158(28):5163–8.
61. Alamili M, Bendtzen K, Lykkesfeldt J, Rosenberg J, Gögenur I. Pronounced inflammatory response to endotoxaemia during nighttime: A randomised cross-over trial. PLoS One. 2014;9(1)
62. Hermann C, von Aulock S, Dehus O, Keller M, Okigami H, Gantner F, et al. Endogenous cortisol determines the circadian rhythm of lipopolysaccharide- but not lipoteichoic acid-inducible cytokine release. Eur J Immunol. 2006;36(2):371–9.
63. Gibbs JE, Blaikley J, Beesley S, Matthews L, Simpson KD, Boyce SH, et al. The nuclear receptor REV-ERBα mediates circadian regulation of innate immunity through selective regulation of inflammatory cytokines. Proc Natl Acad Sci U S A. 2012;109(2):582–7.
64. Spengler ML, Kuropatwinski KK, Comas M, Gasparian AV, Fedtsova N, Gleiberman AS, et al. Core circadian protein CLOCK is a positive regulator of NF-κB-mediated transcription. Proc Natl Acad Sci U S A. 2012;109(37)
65. Early JO, Menon D, Wyse CA, Cervantes-Silva MP, Zaslona Z, Carroll RG, et al. Circadian clock protein BMAL1 regulates IL-1β in macrophages via NRF2. Proc Natl Acad Sci U S A. 2018;115(36):E8460–8.
66. Lee J, Kim DE, Griffin P, Sheehan PW, Kim DH, Musiek ES, et al. Inhibition of REV-ERBs stimulates microglial amyloid-beta clearance and reduces amyloid plaque deposition in the 5XFAD mouse model of Alzheimer's disease. Aging Cell. 2020;19(2):1–15.
67. Kitchen GB, Cunningham PS, Poolman TM, Iqbal M, Maidstone R, Baxter M, et al. The clock gene Bmal1 inhibits macrophage motility, phagocytosis, and impairs defense against pneumonia. Proc Natl Acad Sci U S A. 2020;117(3):1543–51.
68. Huo M, Huang Y, Qu D, Zhang H, Wong WT, Chawla A, et al. Myeloid Bmal1 deletion increases monocyte recruitment and worsens atherosclerosis. FASEB J. 2017;31:1097–106.
69. Yang G, Zhang J, Jiang T, Monslow J, Tang SY, Todd L, et al. Bmal1 deletion in myeloid cells attenuates atherosclerotic lesion development and restrains abdominal aortic aneurysm formation in hyperlipidemic mice. Arterioscler Thromb Vasc Biol. 2020;40:1523–32.
70. Hayashi M, Shimba S, Tezuka M. Characterization of the molecular clock in mouse peritoneal macrophages. Biol Pharm Bull. 2007;30(4):621–6.
71. Bellet MM, Deriu E, Liu JZ, Grimaldi B, Blaschitz C, Zeller M, et al. Circadian clock regulates the host response to Salmonella. Proc Natl Acad Sci U S A. 2013;110(24):9897–902.
72. Silver AC, Arjona A, Walker WE, Fikrig E. The circadian clock controls toll-like receptor 9-mediated innate and adaptive immunity. Immunity. 2012;36(2):251–61.
73. Narasimamurthy R, Hatori M, Nayak SK, Liu F, Panda S, Verma IM. Circadian clock protein cryptochrome regulates the expression of proinflammatory cytokines. Proc Natl Acad Sci U S A. 2012;109(31):12662–7.
74. Sato S, Sakurai T, Ogasawara J, Takahashi M, Izawa T, Imaizumi K, et al. A circadian clock gene, Rev-erbα, modulates the inflammatory function of macrophages through the negative regulation of Ccl2 expression. J Immunol [Internet]. 2014;192(1):407–17. Available from: https://doi.org/10.4049/jimmunol.1301982
75. Ivanov II, McKenzie BS, Zhou L, Tadokoro CE, Lepelley A, Lafaille JJ, et al. The orphan nuclear receptor RORγt directs the differentiation program of proinflammatory IL-17+ T helper cells. Cell. 2006;126(6):1121–33.

76. Makhija S, Griffett JD, Veerakanellore GB, Burris TP, Elgendy B, Griffett K. REV-ERB activation as a novel pharmacological approach for treating inflammatory pain. Front Pharmacol. 2023;14(April):1–10.
77. Chang C, Loo CS, Zhao X, Solt LA, Liang Y, Bapat SP, et al. The nuclear receptor REV-ERBA modulates Th17 cell-mediated autoimmune disease. Proc Natl Acad Sci U S A. 2019;116(37):18528–36.
78. Mia S, Sonkar R, Williams L, Latimer MN, Rawnsley DR, Rana S, et al. Novel roles for the transcriptional repressor E4BP4 in both cardiac physiology and pathophysiology. JACC Basic Transl Sci. 2023;8(9):1141–56.
79. Ince LM, Barnoud C, Lutes LK, Pick R, Wang C, Sinturel F, et al. Influence of circadian clocks on adaptive immunity and vaccination responses. Nat Commun. 2023;14(1)
80. Huang H, Mehta A, Kalmanovich J, Anand A, Bejarano MC, Garg T, et al. Immunological and inflammatory effects of infectious diseases in circadian rhythm disruption and future therapeutic directions. Mol Biol Rep [Internet]. 2023;50(4):3739–53. Available from: https://doi.org/10.1007/s11033-023-08276-w
81. Fonken LK, Frank MG, Kitt MM, Barrientos RM, Watkins LR, Maier SF. Microglia inflammatory responses are controlled by an intrinsic circadian clock. Brain Behav Immun [Internet]. 2015;45:171–9. Available from: https://doi.org/10.1016/j.bbi.2014.11.009
82. Griffin P, Sheehan PW, Dimitry JM, Guo C, Kanan MF, Lee J, et al. Rev-erbα mediates complement expression and diurnal regulation of microglial synaptic phagocytosis. Elife. 2020;9:1–17.
83. Ali AAH, Schwarz-Herzke B, Rollenhagen A, Anstötz M, Holub M, Lübke J, et al. Bmal1-deficiency affects glial synaptic coverage of the hippocampal mossy fiber synapse and the actin cytoskeleton in astrocytes. Glia. 2020;68:947–62.
84. de Miguel M, Kraychete CD, Nascimento JMR. Chronic pain: cytokines, lymphocytes and chemokines [Internet]. Inflamm Allergy Drug Targets (Discontinued). 2014;13:339–49. Available from: http://www.eurekaselect.com/article/64546
85. Grace PM, Rolan PE, Hutchinson MR. Peripheral immune contributions to the maintenance of central glial activation underlying neuropathic pain. Brain Behav Immun [Internet]. 2011;25(7):1322–32. Available from: https://www.sciencedirect.com/science/article/pii/S088915911100119X
86. Sorge RE, Mapplebeck JCS, Rosen S, Beggs S, Taves S, Alexander JK, et al. Different immune cells mediate mechanical pain hypersensitivity in male and female mice. Nat Neurosci. 2016;18(8):1081–3.
87. Mizutani H, Tamagawa-Mineoka R, Minami Y, Yagita K, Katoh N. Constant light exposure impairs immune tolerance development in mice. J Dermatol Sci [Internet]. 2017;86(1):63–70. Available from: https://doi.org/10.1016/j.jdermsci.2016.12.016
88. Bollinger T, Leutz A, Leliavski A, Skrum L, Kovac J, Bonacina L, et al. Circadian clocks in mouse and human CD4+ T cells. PLoS One. 2011;6(12):1–11.
89. Fortier EE, Rooney J, Dardente H, Hardy MP, Labrecque N, Cermakian N. Circadian variation of the response of T cells to antigen. J Immunol [Internet]. 2011;187(12):6291–300. Available from: https://doi.org/10.4049/jimmunol.1004030
90. Kirsch S, Thijssen S, Alarcon Salvador S, Heine GH, Van Bentum K, Fliser D, et al. T-cell numbers and antigen-specific T-cell function follow different circadian rhythms. J Clin Immunol. 2012;32(6):1381–9.
91. Esquifino AI, Chacon F, Cano P, Marcos A, Cutrera RA, Cardinali DP. Twenty-four-hour rhythms of mitogenic responses, lymphocyte subset populations and amino acid content in submaxillary lymph nodes of growing male rats subjected to calorie restriction. J Neuroimmunol [Internet]. 2004;156(1):66–73. Available from: https://www.sciencedirect.com/science/article/pii/S0165572804002310
92. Bollinger T, Bollinger A, Skrum L, Dimitrov S, Lange T, Solbach W. Sleep-dependent activity of T cells and regulatory T cells. Clin Exp Immunol. 2009;155(2):231–8.

93. Yu X, Rollins D, Ruhn KA, Stubblefield JJ, Green CB, Kashiwada M, et al. TH17 cell differentiation is regulated by the circadian clock. Science. 2013;342(6159):727–30.
94. Bahney J, von Bartheld CS, Kwilas AR, Donahue RN, Tsang KY, Hodge JW. The cellular composition and glia–neuron ratio in the spinal cord of a human and a nonhuman primate: comparison with other species and brain regions. Anat Rec. 2018;301(4):1–17.
95. Call CL, Bergles DE. Cortical neurons exhibit diverse myelination patterns that scale between mouse brain regions and regenerate after demyelination. Nat Commun [Internet]. 2021;12(1) Available from: https://doi.org/10.1038/s41467-021-25035-2
96. Saijo K, Glass CK. Microglial cell origin and phenotypes in health and disease. Nat Rev Immunol. 2011;11(11):775–87.
97. Wang S, Li F, Lin Y, Wu B. Targeting REV-ERBα for therapeutic purposes: Promises and challenges. Theranostics. 2020;10(9):4168–82.
98. Wolff SEC, Wang XL, Jiao H, Sun J, Kalsbeek A, Yi CX, et al. The effect of Rev-erbα agonist SR9011 on the immune response and cell metabolism of microglia. Front Immunol. 2020;11(September)
99. Smith JA, Das A, Ray SK, Banik NL. Role of pro-inflammatory cytokines released from microglia in neurodegenerative diseases. Brain Res Bull [Internet]. 2012;87(1):10–20. Available from: https://doi.org/10.1016/j.brainresbull.2011.10.004
100. Brancaccio M, Patton AP, Chesham JE, Maywood ES, Hastings MH. Astrocytes control circadian timekeeping in the suprachiasmatic nucleus via glutamatergic signaling. Neuron [Internet]. 2017;93(6):1420–1435.e5. Available from: https://doi.org/10.1016/j.neuron.2017.02.030
101. Tso CF, Simon T, Greenlaw AC, Puri T, Mieda M, Herzog ED. Astrocytes regulate daily rhythms in the suprachiasmatic nucleus and behavior. Curr Biol [Internet]. 2017;27(7):1055–61. Available from: https://doi.org/10.1016/j.cub.2017.02.037
102. Morioka N, Saeki M, Sugimoto T, Higuchi T, Zhang FF, Nakamura Y, et al. Downregulation of the spinal dorsal horn clock gene Per1 expression leads to mechanical hypersensitivity via c-jun N-terminal kinase and CCL2 production in mice. Mol Cell Neurosci [Internet]. 2016;72:72–83. Available from: https://doi.org/10.1016/j.mcn.2016.01.007
103. Griffin P, Dimitry JM, Sheehan PW, Lananna BV, Guo C, Robinette ML, et al. Circadian clock protein Rev-erb α regulates neuroinflammation. Proc Natl Acad Sci U S A. 2019;116(11):5102–7.
104. Musiek ES, Lim MM, Yang G, Bauer AQ, Qi L, Lee Y, et al. Circadian clock proteins regulate neuronal redox homeostasis and neurodegeneration. J Clin Invest. 2013;123(12):5389–400.
105. Sugimoto T, Morioka N, Zhang FF, Sato K, Abe H, Hisaoka-Nakashima K, et al. Clock gene Per1 regulates the production of CCL2 and interleukin-6 through p38, JNK1 and NF-κB activation in spinal astrocytes. Mol Cell Neurosci [Internet]. 2014;59:37–46. Available from: https://doi.org/10.1016/j.mcn.2014.01.003
106. Wang Y, Jin S, Sonobe Y, Cheng Y, Horiuchi H, Parajuli B, Kawanokuchi J, Mizuno T, Takeuchi H, Suzumura A. Interleukin-1β induces blood–brain barrier disruption by down-regulating sonic hedgehog in astrocytes. PloS one. 2014;9(10):e110024.
107. Clark AK, Old EA, Malcangio M. Neuropathic pain and cytokines: Current perspectives. J Pain Res. 2013;6:803–14.
108. Grace PM, Hutchinson MR, Maier SF, Watkins LR. Pathological pain and the neuroimmune interface. Nat Rev Immunol. 2014;14(4):217–31.
109. Reiter RJ, Sharma R, Cucielo MS, Tan DX, Rosales-Corral S, Gancitano G, et al. Brain washing and neural health: role of age, sleep, and the cerebrospinal fluid melatonin rhythm. Cell Mol Life Sci [Internet]. 2023;80(4):1–22. Available from: https://doi.org/10.1007/s00018-023-04736-5
110. Xu S, Xie X, Yao L, Wang W, Zhang H, Chen M, et al. Human in vivo evidence of reduced astrocyte activation and neuroinflammation in patients with treatment-resistant depression following electroconvulsive therapy. Psychiatry Clin Neurosci. 2023;77(12):653–64.
111. Balsalobre A, Brown SA, Marcacci L, Tronche F, Kellendonk C, Reichardt HM, et al. Resetting of circadian time in peripheral tissues by glucocorticoid signaling. Science. 2000;289(5488):2344–7.

112. Besedovsky L, Born J, Lange T. Endogenous glucocorticoid receptor signaling drives rhythmic changes in human T-cell subset numbers and the expression of the chemokine receptor CXCR4. FASEB J. 2014;28(1):67–75.
113. BaHammam AS, Pirzada A. Timing matters: the interplay between early mealtime, circadian rhythms, gene expression, circadian hormones, and metabolism—a narrative review. Clocks Sleep. 2023;5:507–35.
114. Cheon S, Park N, Cho S, Kim K. Glucocorticoid-mediated Period2 induction delays the phase of circadian rhythm. Nucleic Acids Res. 2013;41(12):6161–74.
115. Fukuoka Y, Burioka N, Takata M, Ohdo S, Miyata M, Endo M, et al. Glucocorticoid administration increases hPer1 mRNA levels in human peripheral blood mononuclear cells in vitro or in vivo [1]. J Biol Rhythms. 2005;20(6):550–3.
116. Rosenfeld P, van Eekelen JAM, Levine S, de Kloet ER. Ontogeny of corticosteroid receptors in the brain. Cell Mol Neurobiol. 1993;13(4):295–319.
117. Segall LA, Amir S. Exogenous corticosterone induces the expression of the clock protein, PERIOD2, in the oval nucleus of the bed nucleus of the stria terminalis and the central nucleus of the amygdala of adrenalectomized and intact rats. J Mol Neurosci. 2010;42(2):176–82.
118. Kiessling S, Eichele G, Oster H. Adrenal glucocorticoids have a key role in circadian resynchronization in a mouse model of jet lag. J Clin Invest. 2010;120(7):2600–9.
119. Saha S, Singh KM, Gupta BBP. Melatonin synthesis and clock gene regulation in the pineal organ of teleost fish compared to mammals: Similarities and differences. Gen Comp Endocrinol [Internet]. 2018 [cited 2018 July 23]. Available from: https://www.sciencedirect.com/science/article/pii/S0016648018302132?via%3Dihub
120. Mauriz JL, Collado PS, Veneroso C, Reiter RJ, González-Gallego J. A review of the molecular aspects of melatonin's anti-inflammatory actions: Recent insights and new perspectives. J Pineal Res. 2013;54(1):1–14.
121. Muxel SM, Pires-Lapa MA, Monteiro AWA, Cecon E, Tamura EK, Floeter-Winter LM, et al. NF-κB drives the synthesis of melatonin in RAW 264.7 macrophages by inducing the transcription of the arylalkylamine-N-acetyltransferase (AA-NAT) Gene. PLoS One. 2012;7(12)
122. Cecon E, Fernandes PA, Pinato L, Ferreira ZS, Markus RP. Daily variation of constitutively activated nuclear factor kappa B (NFKB) in rat pineal gland. Chronobiol Int [Internet]. 2010;27(1):52–67. Available from: https://doi.org/10.3109/07420521003661615
123. Ferreira ZS, Bothorel B, Markus RP, Simonneaux V. Plasma corticosterone elevation inhibits the activation of nuclear factor kappa B (NFKB) in the Syrian hamster pineal gland. Stress. 2012;15(3):339–47.
124. Leng Y, Musiek ES, Hu K, Cappuccio FP, Yaffe K, Francisco S, et al. Association between circadian rhythms and neurodegenerative diseases. Lancet Neurol. 2020;18(3):307–18.
125. Hickman S, Izzy S, Sen P, Morsett L, El Khoury J. Microglia in neurodegeneration. Nat Neurosci [Internet]. 2018;21(10):1359–69. Available from: https://doi.org/10.1038/s41593-018-0242-x
126. Jasinska M, Grzegorczyk A, Woznicka O, Jasek E, Kossut M, Barbacka-Surowiak G, et al. Circadian rhythmicity of synapses in mouse somatosensory cortex. Eur J Neurosci [Internet]. 2015;42(8):2585–94. Available from: https://doi.org/10.1111/ejn.13045
127. Krzeptowski W, Hess G, Pyza E. Circadian plasticity in the brain of insects and rodents. Front Neural Circuits. 2018;12(May):1–14.
128. Hasegawa S, Fukushima H, Hosoda H, Serita T, Ishikawa R, Rokukawa T, et al. Hippocampal clock regulates memory retrieval via dopamine and PKA-induced GluA1 phosphorylation. Nat Commun [Internet]. 2019;10(1):1–14. Available from: https://doi.org/10.1038/s41467-019-13554-y
129. Leng Y, Blackwell T, Cawthon PM, Ancoli-Israel S, Stone KL, Yaffe K, et al. Association of circadian abnormalities in older adults with an increased risk of developing parkinson disease. Science [Internet]. 2019;9(1):1–26. Available from: https://www.nature.com/articles/nature00965

130. Lananna BV, McKee CA, King MW, Del-Aguila JL, Dimitry JM, Farias FHG, et al. Chi3l1/YKL-40 is controlled by the astrocyte circadian clock and regulates neuroinflammation and Alzheimer's disease pathogenesis. Sci Transl Med. 2020;12(574)
131. Jack CR, Bennett DA, Blennow K, Carrillo MC, Dunn B, Haeberlein SB, et al. NIA-AA Research Framework: toward a biological definition of Alzheimer's disease. Alzheimers Dement. 2018;14(4):535–62.
132. Kress GJ, Liao F, Dimitry J, Cedeno MR, FitzGerald GA, Holtzman DM, et al. Regulation of amyloid-β dynamics and pathology by the circadian clock. J Exp Med. 2018;215(4):1059–68.
133. Ni J, Wu Z, Meng J, Saito T, Saido TC, Qing H, et al. An impaired intrinsic microglial clock system induces neuroinflammatory alterations in the early stage of amyloid precursor protein knock-in mouse brain. J Neuroinflamm. 2019;16(1):1–15.
134. Song H, Moon M, Choe HK, Han DH, Jang C, Kim A, et al. Aβ-induced degradation of BMAL1 and CBP leads to circadian rhythm disruption in Alzheimer's disease. Mol Neurodegener. 2015;10(1):1–15.
135. Stevanovic K, Yunus A, Joly-Amado A, Gordon M, Morgan D, Gulick D, et al. Disruption of normal circadian clock function in a mouse model of tauopathy. Exp Neurol [Internet]. 2017;294:58–67. Available from: https://doi.org/10.1016/j.expneurol.2017.04.015
136. Burns C, Bloom C, Said M, Pinto L, Azam S, Flores M, et al. Inhibition of casein kinase 1δ/ε improves cognitive-affective behavior and reduces amyloid load in the APP-PS1 mouse model of Alzheimer's disease. Sci Rep. 2019;9(13743):1–13.
137. Kondratova AA, Dubrovsky YV, Antoch MP, Kondratov RV. Circadian clock proteins control adaptation to novel environment and memory formation. Aging (Albany NY). 2010;2(5):285–97.
138. Pacelli C, Rotundo G, Lecce L, Menga M, Bidollari E, Scrima R, et al. Parkin mutation affects clock gene-dependent energy metabolism. Int J Mol Sci. 2019;20(11)
139. Gu Z, Wang B, Zhang YB, Ding H, Zhang Y, Yu J, et al. Association of ARNTL and PER1 genes with Parkinson's disease: A case-control study of Han Chinese. Sci Rep. 2015;5:1–10.
140. Li H, Song S, Wang Y, Huang C, Zhang F, Liu J, et al. Correction to: Low-grade inflammation aggravates rotenone neurotoxicity and disrupts circadian clock gene expression in rats (Neurotox Res. 2019;35(2):421–31. 10.1007/s12640-018-9968-1). Neurotox Res. 2019;35(4):999–1000.
141. Liu WW, Wei SZ, Huang GD, Liu LB, Gu C, Shen Y, et al. BMAL1 regulation of microglia-mediated neuroinflammation in MPTP-induced Parkinson's disease mouse model. FASEB J. 2020;34:6570–81.
142. Barca-Mayo O, Pons-Espinal M, Follert P, Armirotti A, Berdondini L, De Pietri Tonelli D. Astrocyte deletion of Bmal1 alters daily locomotor activity and cognitive functions via GABA signalling. Nat Commun. 2017;8
143. Farina MG, Sandhu MRS, Parent M, Sanganahalli BG, Derbin M, Dhaher R, et al. Small loci of astroglial glutamine synthetase deficiency in the postnatal brain cause epileptic seizures and impaired functional connectivity. Epilepsia. 2021;62(11):2858–70.
144. Dimitrov S, Benedict C, Heutling D, Westermann J, Born J, Lange T. Cortisol and epinephrine control opposing circadian rhythms in T cell subsets. Blood, the Journal of the American Society of Hematology. 2009;113(21):5134–5143.
145. Carrillo-Vico A, Lardone PJ, Fernández-Santos JM, Martín-Lacave I, Calvo JR, Karasek M, Guerrero JM. Human lymphocyte-synthesized melatonin is involved in the regulation of the interleukin-2/interleukin-2 receptor system. The Journal of Clinical Endocrinology & Metabolism. 2005;90(2):992–1000.

Circadian Regulation of Peripheral Inflammation

Ashutosh K. Tiwari and Ratnasekhar Ch

Abstract Circadian rhythms regulate various physiological and cellular processes. These biological oscillations are fundamental features of nearly all organisms. In mammals, light synchronizes brain clocks, while feeding is the primary cause of peripheral clock synchronization in metabolic tissues such as the muscles, adipose tissues, liver, and pancreas. The circadian clock regulates several intracellular signaling pathways, including inflammatory and immunological response, angiogenesis, metabolic and redox homeostasis, and cell proliferation. Furthermore, circadian dysfunction caused by the misalignment of peripheral clocks leads to various diseases such as metabolic disorders, neurodegeneration, cancer, and cardiovascular disease. The underlying molecular mechanism of the circadian clock is linked to peripheral inflammation, activation of the innate or adaptive immune system, and the release of proinflammatory cytokines in response to a variety of pathogenic stimuli. In this book chapter, we emphasize the circadian clock mechanism and their regulation of peripheral inflammation and associated immunological disorders.

Keywords Circadian clock · Peripheral inflammation · Proinflammatory cytokines
Immunological disorder

A. K. Tiwari
Metabolomics Lab, CSIR-Central Institute of Medicinal & Aromatic Plants (CIMAP), Lucknow, Uttar Pradesh, India

Ratnasekhar Ch (✉)
Metabolomics Lab, CSIR-Central Institute of Medicinal & Aromatic Plants (CIMAP), Lucknow, Uttar Pradesh, India

Academy of Council of Scientific and Industrial Research (AcSIR), Ghaziabad, Uttar Pradesh, India
e-mail: ratnasekhar@cimap.res.in

P. K. Jha, H. Bouaouda (eds.), *Circadian Rhythms, Sleep and Inflammation*, Progress in Inflammation Research 93,
https://doi.org/10.1007/978-3-031-74744-1_2

1 Introduction

The circadian rhythm (Latin *circa diem*, for about a day), 24 h endogenous biological oscillation, is a fundamental feature of virtually all organisms and regulates the physiological and cellular processes including inflammation [1, 2]. The alignment of an endogenous biological clock to an external environment is called entrainment. Light is the crucial external cues that helps to maintaine an organism's internal daily clock with natural cycle of day and night. Furthermore, indirectly it influences the timing of eating, another potent entrainer of circadian rhythm [3]. In mammals, circadian system is composed of central clock which is located in the suprachiasmatic nucleus (SCN) of the hypothalamus and peripheral clocks are dispersed throughout the body. The peripheral clock is composed of metabolic organs including the liver, pancreas, muscles, and adipose tissues. Evidence of circadian oscillations in immune system components has grown over the past 15 years, indicating that these oscillations are a crucial regulator of immunological components and have the potential to influence disease onset and treatments [4]. The circadian clock system composed of autoregulatory transcription-translation feedback loops that are provided by several sets of transcription factors. Circadian clock genes include BMAL1 (brain and muscle aryl hydrocarbon receptor nuclear translocator (ARNT)-like 1), CLOCK (circadian locomotor output cycles kaput), and the CLOCK-related gene NPAS2 (neuronal PAS domain-containing protein 2, expressed in the forebrain). Clock genes, i.e., *CLOCK* and *BMAL1*, combine to form the heterodimer complex in the cytoplasm, which translocates in the nucleus where it binds to the E-box and E′-box of the promoter region of the clock-controlled genes [1, 2, 5].

In circadian clock system, *CLOCK* and *BMAL1* proteins induce the production of their opposite regulators, *CRY* (cryptochrome) and *PER* (period circadian protein), which in turn promotes their own repression. The heterodimer complex formed by the negative regulators PER and CRY is translocated to the nucleus where it interacts with BMAL1-CLOCK dimer to prevent transcription. The decreased expression of PER and CRY associated with lowered BMAL1-CLOCK levels and then restart the natural clock cycle. An additional feedback loop i.e. nuclear receptor REV-ERB (encoded by NR1D1) stabilizes the circadian clock that make up a larger portion of the clock loop that capable of governing circadian rhythms (Fig. 1) [3, 6]. These clock events are responsible for the regulation of immunological functions.

1.1 Clock Rhythmicity in Immunological Parameter

There are several immunological parameters including eosinophils, mast cells, neutrophils, monocytes, macrophages, and natural killer (NK) cells and different hormones or cytokines exhibit intrinsic clock rhythmicity [7–9]. The intrinsic clock can have an impact on how these cells function, which controls the temporal gating of

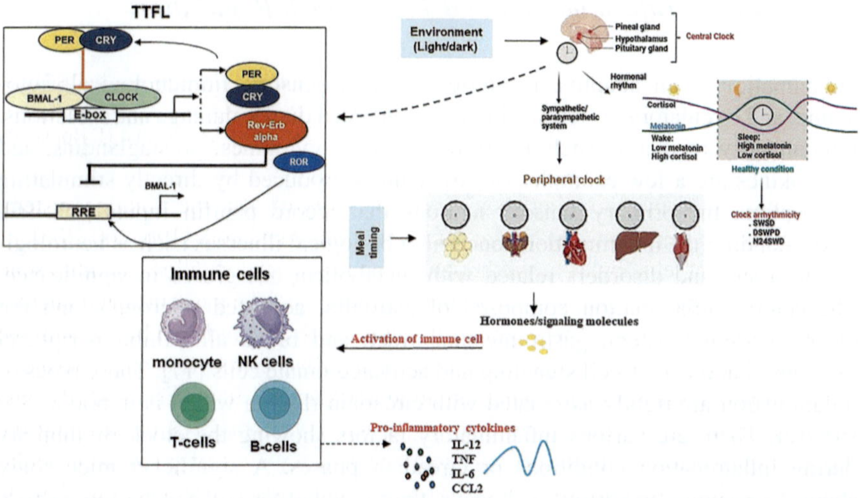

Fig. 1 Schematic representation of the regulation of central and peripheral clocks on immune response. The mammalian circadian system consists of central and peripheral clocks. The central clock, located in the suprachiasmatic nucleus (SCN) of the brain, is regulated by environmental cues, while peripheral clocks are influenced by factors such as meal intake. The SCN communicates with peripheral clocks in various tissues—including the liver, cardiovascular system, kidney, adipose tissue, and muscles—through both synaptic and diffusible signals, orchestrating daily patterns of behavior and physiological functions. Additionally, peripheral tissues have their intrinsic clocks that can interact with each other. The BMAL1: CLOCK heterodimer complex binds to the E-box regions of target genes to initiate transcription. Cytoplasmic PER and CRY proteins form dimers, translocate to the nucleus, and inhibit transcription of their target genes. For the circadian clock to maintain its rhythmicity and start a new cycle, the degradation of these negative feedback loop proteins, PER and CRY, is essential after the repression phase of the transcriptional-translational feedback loop (TTFL) concludes. The peripheral clock enables to release of some signaling molecules to activate the immune cells which release proinflammatory molecules that regulate the immunological function with clock rhythmicity. Clock arrhythmicity; *SWSD* shift work sleep disorder, *DSWPD* delayed sleep-wake phase disorder, *N24SWD* non-24-h sleep-wake rhythm disorder (created with biorender.com, PowerPoint presentation)

cell type-specific events. The temporal gating encompasses the histamine release, mast cell allergic reaction, phagocytic activity, and cytokine release from macrophages [10, 11]. There are limited studies suggest that specific clock genes that mediate different cellular physiology of immune system. One of the study revealed the importance of the clock gene PER1 which is indispensable for the proper functioning of NK cells in the splenic system [12]. However, these cells still exhibit rhythmicity in the absence of Per1, and exhibit altered rhythmic production of interferon (IFN), as well as the cytotoxic agent's perforin and granzyme B. Additionally, chronic jet lag can also suppress the clock rhythmicity in these components and the related cytotoxicity [13]. Furthermore, cryptochromes (CRYs) can affect the inflammatory pathways through the action of adenyl cyclase. Absence of cryptochrome proteins constitutively activates the proinflammatory cytokines expression [14]. These proteins are responsible for the regulation of the various pathophysiological functions.

1.2 Types of Inflammation and Their Clock Regulation

Inflammation is an adaptive response of somatosensory, immunological, autonomic, and circulatory systems, which are generated during damage and infections. Inflammatory mediators such as proinflammatory cytokines, prostaglandins, and chemokines are a few examples of how pain is produced by directly stimulating nociceptors, the primary sensory neurons that record painful inputs [15, 16]. Understanding the inflammation concerning biological illnesses such as neurological disorders and disorders related with metabolism has grown in significance. The central inflammation composed of astroglia, activated microglia, and the enteric nervous system, gastrointestinal tract, and blood all exhibit peripheral inflammation due to T cell signaling and activated innate cells [17]. These types of inflammation are tightly associated with circadian rhythm which is important for survival. There are various inflammatory factors showing the clock rhythmicity during inflammation conditions or circadian phases. A significant mice study showed a response to various pathogens, their metabolites (such as bacterial endotoxins and exotoxins), and proinflammatory cytokines that are regulated by the circadian rhythm [18, 19]. Collectively, these investigations demonstrated mice that are more vulnerable to the substances during the beginning of the active phase or in the early evening. The circadian exacerbations in the symptoms of some chronic diseases are well established. For example, early morning joint stiffness is a symptom of rheumatoid arthritis and has been linked to elevated serum levels of TNF and IL-6 [20]. At this period, i.e., early morning, people are most susceptible to cardiovascular problems including myocardial infarction and stroke [21, 22]. This has been associated to an increase in sympathetic nervous system activity in the early morning, which causes the blood to become more viscous and coagulable and increases the risk of thrombosis [23]. The fact that vaso-occlusive events are affected by the time of day in both mice [24] and humans [25] is further evidence that these events contribute to circadian fluctuations in the onset of vascular ailments. Additionally, recent studies suggest that ischemic brain cells can release a protein such as peroxiredoxins are antioxidant proteins found in the cytoplasm that are necessary for maintaining redox equilibrium and act as conserved biomarkers of clock rhythmicity [26]. These proteins can bind to toll-like receptor 2 (TLR2) and toll-like receptor 4 (TLR4) in brain-invading macrophages resulting in the production of interleukin 23 (IL23) and causing the attraction of proinflammatory IL17+ T cells, both of which may exacerbate brain injury [27]. It is believed that a variety of cellular and humoral processes controlled by the circadian clock influence the onset and development of disease.

2 Role of Circadian Clock Rhythmicity in Distinct Tissue Inflammation

Every cell in the body has a circadian oscillator, which is managed by a master clock and gives specific cells and organs rhythmicity by managing rate-limiting metabolic processes [28]. It is well-known that various metabolite levels oscillate in tissues and plasma, and there is mounting proof that disruptions in the circadian clock expression have a causative role in the onset of systemic diseases [29].

2.1 Adipose Tissues

It is believed to be primarily involved in energy storage; white adipose tissue is now widely recognized to have an endocrine function and release several substances known as adipokines [30]. White adipose tissue experiences sterile inflammation as a result of the metabolic stress brought on by obesity. This sterile inflammation encourages a shift in the synthesis of proinflammatory adipokines, which aids in the onset of diabetes and systemic metabolic dysfunction [31]. One of the important studies demonstrated that AP2-Cre Bmal1fl/fl mice with a C57BL/6J genetic background have been shown to exhibit increased food intake and acquire obesity when given a regular chow diet. When the adipocyte clock is perturbed, the levels of polyunsaturated fatty acids in the blood fluctuate throughout time, which causes variations in the expression of neurotransmitters that control hunger in the hypothalamic feeding center [32, 33]. It is commonly recognized that the development of systemic metabolic dysfunction is facilitated by the abnormal expansion of sterile chronic inflammation in white adipose tissue. Clock genes are essential to control how immune cells respond. A significant studies conducted on Per1 and Per2 are specifically disrupted in myeloid cells, which has been linked to increased body weight, visceral adipose tissue weight, dietary adipose tissue inflammation, and insulin resistance [34]. In addition to these, one of the important studies revealed that the obesity and metabolic inflammation are linked by circadian clock malfunction in human omental fat. They found a connection between NF-κB and the adipose tissue's molecular clock that may be responsible for the difficulties associated with obesity [35].

2.2 Skeletal Muscles

The skeletal muscle system has a circadian clock system, and its perturbation is associated with alterations in the muscles' phenotype and function. The anabolism and catabolism must coexist in equilibrium for the maintenance of muscle mass. A worse quality of life, fragility, and a shorter lifespan are all strongly associated with skeletal muscle atrophy (loss of skeletal muscle mass) [36]. Sarcopenia, or age-related muscle loss, Duchenne muscular dystrophy (atrophy caused by a lack of dystrophin), cancer cachexia, disuse-induced atrophy (i.e., extended bed rest), and other conditions are examples of atrophic conditions [37]. These conditions are characterized by circadian dysfunction and chronic systemic elevation of inflammatory cytokine. Studies on the relationship between inflammatory cytokines and the circadian clock in skeletal muscle atrophy are rare. However, there is proof that the expression of cytokines, including IL-6 and IL-8 in human skeletal muscle cells, has a diurnal rhythm in skeletal muscle and other cell types. IL-6 and IL-8 secretion in myotubes derived from humans is significantly reduced when *Clock* gene is silenced demonstrating a clear relationship between the skeletal muscle clock and cytokine release [38]. When Cry1/2 is deleted, whole body exhibits a severe inflammatory phenotype with increased splenocyte counts and blood levels of TNF and IL-6, which exacerbates arthritic symptoms. These findings imply that clock genes are involved in the regulation of inflammatory and immunological responses [39].

2.3 Liver

It is the largest gland in the human body and is vital to the regulation of systemic metabolism due to its involvement in protein synthesis, hormone production, glycogen storage, detoxification, and immune system control [40]. The pharmacokinetics of xenobiotics is tightly regulated by the liver's circadian clock during multiple phases, such as uptake, distribution, metabolism, elimination, absorption, and clearance. Misalignment of the circadian clock in the liver results in inflammations including fatty liver, cholestasis, hepatitis, cirrhosis, and liver cancer, all of which can disrupt the clock rhythmicity [41, 42]. Non-alcoholic fatty liver disease (NAFLD) is the accumulation of triglyceride (TG) at or above 5% of liver weight without usage of alcohol. The metabolism of liver lipids is strictly regulated by circadian clock genes, and liver lipid accumulation is caused by a disruption of the circadian rhythms [43, 44]. *Bmal1* is crucial in the regulation of fat storage, utilization, and adipocyte differentiation. One of the significant studies revealed the deletion of Bmal1 in animals results in reduced fat storage, hepatic steatosis, hypoinsulinemia, higher levels of circulating fatty acids, significant fat accumulation in the liver and muscle, and intolerance of glucose [45]. Similarly, the Clock is also associated with NAFLD. One of the significant studies revealed a reduced hepatic triglyceride accumulation under high-fat diet (HFD) conditions in Clock knockout mice. Obesity, steatosis, hyperglycemia, and hypertriglyceridemia are all distinct characteristics of clock double mutant mice, who also exhibit enhanced lipid absorption throughout the day [46, 47].

2.4 Pancreas

It is a multifunctional organ that controls digestion and metabolism through several endocrine and exocrine processes. These tightly associated processes are crucial for glucose homeostasis, food intake, and lipolysis [48, 49]. Numerous cell types and hormone releases are influenced by circadian rhythms, and both distal (the liver, gut, and brain) and local (the pancreas) circadian rhythms have a significant effect on insulin resistance and blood glucose homeostasis [50, 51]. Important hormones like glucagon and insulin exhibit circadian rhythmicity, and type 2 diabetes is brought on by their arrhythmia. Insulin receptor deterioration and dysregulated glucose metabolism are hallmarks of type 2 diabetes mellitus [52]. The molecular clock mechanism is very important to understand the pancreas physiology. One of the important studies conducted on *bmal1* deficient mice revealed to development hyperglycemia and hypoinsulinemia which is the characteristic of diabetes [53]. One of the most common inflammations is pancreatitis which is the result of dysregulated circadian clock. In pancreatitis, initially, pancreatic acinar cells, which are found inside the (typically sterile) pancreas, are damaged. When there is damage, intracellular pancreatic enzymes such as amylase, lipase, and proteases are mistakenly released from the acinar cells and cause damage to the neighboring healthy pancreatic tissue [54, 55]. Both the deletion of BMAL1 and the CLOCK gene result in a significant loss of islets of Langerhans, poor glucose tolerance, and impaired insulin synthesis, all of which have an impact on glucagon secretion. This irregular circadian gene expression in peripheral tissues as well as the SCN may thus have an adverse effect on pancreatitis via dysregulation of the immune response [53, 56]. The pathogenic mechanisms underlying and systemic inflammation caused by other conditions share a lot of similarities.

2.5 Cardiovascular System

The cardiovascular system is composed of the heart and it is responsible for transporting blood through vessels to and from every region of the body, supplying tissues with nutrition and oxygen, and eliminating waste products such as carbon dioxide and other toxins [57]. It is now understood that the circadian clock regulates several cardiovascular processes, including blood pressure, heart rate, thrombus development, and endothelial function. Circadian arrhythmia causes cardiovascular inflammation or disorder, myocardial infarction (MI), and stroke that is associated with circadian rhythmicity [58]. There are many MI risk factors, such as high blood pressure, which have a 24-h cycle; interference with these physiological rhythms is linked to an increased risk of MI [59]. The risk of MI is increased by metabolic syndrome, hypertension, and diabetes mellitus, all of which are related to single nucleotide polymorphisms in clock genes [60]. When an MI occurs in the morning, neutrophil recruitment to wounded tissues is at its peak, which causes greater inflammation. Pharmacological circadian modulators have been shown to partially mitigate the negative effects of myocardial remodeling in animal models after MI, which is related to disturbance of the circadian clock [61, 62].

3 Conclusion

Every day physiological processes of individuals are adjusted by their circadian rhythm. Numerous studies have provided strong connections between clock rhythmicity and immunological functions. Clock rhythmicity is indispensable for the regulation of various inflammations. However, but there is still a lot of effort to be done to put research into practice in the clinic and integrate the variety of circadian effects into a complicated immune response. To find novel opportunities in the creation of time-based interrelations, it will be essential to have a thorough grasp of the molecular mechanisms underlying circadian modulation of peripheral inflammation.

Compliance with Ethical Standards The authors declare no conflict of interest. No ethical approval was required for this study.

Funding The authors would like to thank the Department of Science and Technology (DST), Government of India for the SERB's SRG research grant (SRG/2021/000750-G), Department of Biotechnology, Government of India for the Ramalingaswami grant (BT/RLF/Re-entry/21/2020), for funding the present work.

References

1. Scheiermann C, Kunisaki Y, Frenette PS. Circadian control of the immune system. Nat Rev Immunol. 2013;13(3):190–8. https://doi.org/10.1038/nri3386.
2. Tiwari A, Rathor P, Trivedi PK, Ch R. Multi-omics reveal interplay between circadian dysfunction and type2 diabetes. Biology. 2023;12(2):301. https://doi.org/10.3390/biology12020301.
3. Golombek DA, Rosenstein RE. Physiology of circadian entrainment. Physiol Rev. 2010;90(3):1063–102. https://doi.org/10.1152/physrev.00009.2009.
4. Lange T, Dimitrov S, Born J. Effects of sleep and circadian rhythm on the human immune system. Ann N Y Acad Sci. 2010;1193:48–59. https://doi.org/10.1111/j.1749-6632.2009.05300.x.
5. Radcliffe LA, Bradfield CA, Clendenin C, Simon MC, Moran SM, Hogenesch JB, Takahashi JS, Wilsbacher LD, Bunger MK. Mop3 is an essential component of the master circadian pacemaker in mammals. Cell. 2000;103(7):1009–17.
6. Cho H, Zhao X, Hatori M, Yu RT, Barish GD, Lam MT, Chong LW, Ditacchio L, Atkins AR, Glass CK, Liddle C, Auwerx J, Downes M, Panda S, Evans RM. Regulation of circadian behaviour and metabolism by REV-ERB-α and REV-ERB-β. Nature. 2012;485:123–7. https://doi.org/10.1038/nature11048.
7. Nguyen KD, Fentress SJ, Qiu Y, Yun K, Cox JS, Chawla A. Circadian gene bmal1 regulates diurnal oscillations of Ly6Chi inflammatory monocytes. Science. 2013;341(6153):1483–8. https://doi.org/10.1126/science.1240636.

8. Keller M, Mazuch J, Abraham U, Eom GD, Herzog ED, Volk HD, Kramer A, Maier B. A circadian clock in macrophages controls inflammatory immune responses. Proc Natl Acad Sci USA. 2009;106(50):21407–12. https://doi.org/10.1073/pnas.0906361106.
9. Arjona A, Sarkar DK. Circadian oscillations of clock genes, cytolytic factors, and cytokines in rat NK cells. J Immunol. 2005;174(12):7618–24. https://doi.org/10.4049/jimmunol.174.12.7618.
10. Nakamura Y, Nakano N, Ishimaru K, Hara M, Ikegami T, Tahara Y, Katoh R, Ogawa H, Okumura K, Shibata S, Nishiyama C, Nakao A. Circadian regulation of allergic reactions by the mast cell clock in mice. J Allergy Clin Immunol. 2014;133(2):568–575.e12. https://doi.org/10.1016/j.jaci.2013.07.040.
11. Nakamura Y, Ishimaru K, Shibata S, Nakao A. Regulation of plasma histamine levels by the mast cell clock and its modulation by stress. Sci Rep. 2017;7(Sept 2016):1–12. https://doi.org/10.1038/srep39934.
12. Logan RW, Wynne O, Levitt D, Price D, Sarkar DK. Altered circadian expression of cytokines and cytolytic factors in splenic natural killer cells of Per1-/- mutant mice. J Interf Cytokine Res. 2013;33(3):108–14. https://doi.org/10.1089/jir.2012.0092.
13. Logan RW, Zhang C, Murugan S, O'Connell S, Levitt D, Rosenwasser AM, Sarkar DK. Chronic shift-lag alters the circadian clock of NK cells and promotes lung cancer growth in rats. J Immunol. 2012;188(6):2583–91. https://doi.org/10.4049/jimmunol.1102715.
14. Narasimamurthy R, Hatori M, Nayak SK, Liu F, Panda S, Verma IM. Circadian clock protein cryptochrome regulates the expression of proinflammatory cytokines. Proc Natl Acad Sci USA. 2012;109(31):12662–7. https://doi.org/10.1073/pnas.1209965109.
15. Matsuda M, Huh Y, Ji RR. Roles of inflammation, neurogenic inflammation, and neuroinflammation in pain. J Anesth. 2019;33(1):131–9. https://doi.org/10.1007/s00540-018-2579-4.
16. Barik S. Molecular interactions between pathogens and the circadian clock. Int J Mol Sci. 2019;20(23) https://doi.org/10.3390/ijms20235824.
17. Williams GP, Schonhoff AM, Sette A, Lindestam Arlehamn CS. Central and peripheral inflammation: connecting the immune responses of Parkinson's disease. J Parkinsons Dis. 2022;12:S129–36. https://doi.org/10.3233/JPD-223241.
18. Shackelford PG, Feigin RD. Periodicity of susceptibility to pneumococcal infection: influence of light and adrenocortical secretions. Science. 1973;182(4109):285–7. Published by American Association for the Advancement of Science
19. House SD, Ruth S, Iii WFK, Rocholl CW, Moldow RL, Ave SO, Orange S. Effects of the circadian rhythm of corticosteroids on leukocyte-endothelium interactions in the AM and PM. Life Sci. 1997;60(22):2023–34.
20. Buttgereit F, Smolen JS, Coogan AN, Cajochen C. Clocking in: chronobiology in rheumatoid arthritis. Nat Rev Rheumatol. 2015;11(6):349–56. https://doi.org/10.1038/nrrheum.2015.31.
21. Thompson DR, Pohl JEF, Sutton TW. Circadian variation in the frequency of onset of chest pain in elderly patients with acute myocardial infarction. Age Ageing. 1992;21(2):99–102. https://doi.org/10.1093/ageing/21.2.99.
22. Jeyaraj D, Haldar SM, Wan X, McCauley MD, Ripperger JA, Hu K, Lu Y, Eapen BL, Sharma N, Ficker E, Cutler MJ, Gulick J, Sanbe A, Robbins J, Demolombe S, Kondratov RV, Shea SA, Albrecht U, Wehrens XHT, Rosenbaum DS, Jain MK. Circadian rhythms govern cardiac repolarization and arrhythmogenesis. Nature. 2012;483(7387):96–101. https://doi.org/10.1038/nature10852.
23. Marfella R, Siniscalchi M, Portoghese M, Di Filippo C, Ferraraccio F, Schiattarella C, Crescenzi B, Sangiuolo P, Ferraro G, Siciliano S, Cinone F, Mazzarella G, Martis S, Verza M, Coppola L, Rossi F, D'Amico M, Paolisso G. Morning blood pressure surge as a destabilizing factor of atherosclerotic plaque: role of ubiquitin-proteasome activity. Hypertension. 2007;49(4):784–91. https://doi.org/10.1161/01.HYP.0000259739.64834.d4.
24. Scheiermann C, Kunisaki Y, Lucas D, Chow A, Jang JE, Zhang D, Hashimoto D, Merad M, Frenette PS. Adrenergic nerves govern circadian leukocyte recruitment to tissues. Immunity. 2012;37(2):290–301. https://doi.org/10.1016/j.immuni.2012.05.021.

25. Auvil-Novak SE, Novak RD, El Sanadi N. Twenty-four-hour pattern in emergency department presentation for sickle cell vaso-occlusive pain crisis. Chronobiol Int. 1996;13(6):449–56. https://doi.org/10.3109/07420529609020915.
26. Edgar RS, Green EW, Zhao Y, Van Ooijen G, Olmedo M, Qin X, Xu Y, Pan M, Valekunja UK, Feeney KA, Maywood ES, Hastings MH, Baliga NS, Merrow M, Millar AJ, Johnson CH, Kyriacou CP, O'Neill JS, Reddy AB. Peroxiredoxins are conserved markers of circadian rhythms. Nature. 2012;485(7399):459–64. https://doi.org/10.1038/nature11088.
27. Shichita T, Hasegawa E, Kimura A, Morita R, Sakaguchi R, Takada I, Sekiya T, Ooboshi H, Kitazono T, Yanagawa T, Ishii T, Takahashi H, Mori S, Nishibori M, Kuroda K, Akira S, Miyake K, Yoshimura A. Peroxiredoxin family proteins are key initiators of post-ischemic inflammation in the brain. Nat Med. 2012;18(6):911–7. https://doi.org/10.1038/nm.2749.
28. Reinke H. Crosstalk between metabolism and circadian clocks. Nat Rev Mol Cell Biol. 2019; https://doi.org/10.1038/s41580-018-0096-9.
29. Shimizu I, Yoshida Y, Minamino T. A role for circadian clock in metabolic disease. Hypertens Res. 2016;39(7):483–91. https://doi.org/10.1038/hr.2016.12.
30. Zvonic S, Ptitsyn AA, Conrad SA, Scott LK, Floyd ZE, Kilroy G, Wu X, Goh BC, Mynatt RL, Gimble JM. Characterization of peripheral circadian clocks in adipose tissues. Diabetes. 2006;55(4):962–70. https://doi.org/10.2337/diabetes.55.04.06.db05-0873.
31. Ando H, Yanagihara H, Hayashi Y, Obi Y, Tsuruoka S, Takamura T, Kaneko S, Fujimura A. Rhythmic messenger ribonucleic acid expression of clock genes and adipocytokines in mouse visceral adipose tissue. Endocrinology. 2005;146(12):5631–6. https://doi.org/10.1210/en.2005-0771.
32. Prasai MJ, Mughal RS, Wheatcroft SB, Kearney MT, Grant PJ, Scott EM. Diurnal variation in vascular and metabolic function in diet-induced obesity: divergence of insulin resistance and loss of clock rhythm. Diabetes. 2013;62(6):1981–9. https://doi.org/10.2337/db11-1740.
33. Paschos GK, Ibrahim S, Song WL, Kunieda T, Grant G, Reyes TM, Bradfield CA, Vaughan CH, Eiden M, Masoodi M, Griffin JL, Wang F, Lawson JA, Fitzgerald GA. Obesity in mice with adipocyte-specific deletion of clock component Arntl. Nat Med. 2012;18(12):1768–77. https://doi.org/10.1038/nm.2979.
34. Xu H, Li H, Woo SL, Kim SM, Shende VR, Neuendorff N, Guo X, Guo T, Qi T, Pei Y, Zhao Y, Hu X, Zhao J, Chen L, Chen L, Ji JY, Alaniz RC, Earnest DJ, Wu C. Myeloid cell-specific disruption of period1 and period2 exacerbates diet-induced inflammation and insulin resistance. J Biol Chem. 2014;289(23):16374–88. https://doi.org/10.1074/jbc.M113.539601.
35. Maury E, Navez B, Brichard SM. Circadian clock dysfunction in human omental fat links obesity to metabolic inflammation. Nat Commun. 2021;12(1):1–17. https://doi.org/10.1038/s41467-021-22571-9.
36. Zhang H, Liang J, Chen N. Do not neglect the role of circadian rhythm in muscle atrophy. Ageing Res Rev. 2020;63(May):101155. https://doi.org/10.1016/j.arr.2020.101155.
37. Baltgalvis KA, Berger FG, Pena MMO, Davis JM, Muga SJ, Carson JA. Interleukin-6 and cachexia in ApcMin/+ mice. Am J Physiol Regul Integr Comp Physiol. 2008;294(2):1–35. https://doi.org/10.1152/ajpregu.00716.2007.
38. Perrin L, Loizides-Mangold U, Skarupelova S, Pulimeno P, Chanon S, Robert M, Bouzakri K, Modoux C, Roux-Lombard P, Vidal H, Lefai E, Dibner C. Human skeletal myotubes display a cell-autonomous circadian clock implicated in basal myokine secretion. Mol Metab. 2015;4(11):834–45. https://doi.org/10.1016/j.molmet.2015.07.009.
39. Hashiramoto A, Yamane T, Tsumiyama K, Yoshida K, Komai K, Yamada H, Yamazaki F, Doi M, Okamura H, Shiozawa S. Mammalian clock gene cryptochrome regulates arthritis via proinflammatory cytokine TNF-α. J Immunol. 2010;184(3):1560–5. https://doi.org/10.4049/jimmunol.0903284.
40. Ferrarini A, Di Poto C, He S, Tu C, Varghese RS, Kara Balla A, Jayatilake M, Li Z, Ghaffari K, Fan Z, Sherif ZA, Kumar D, Kroemer A, Tadesse MG, Ressom HW. Metabolomic analysis of liver tissues for characterization of hepatocellular carcinoma. J Proteome Res. 2019;18(8):3067–76. https://doi.org/10.1021/acs.jproteome.9b00185.

41. McCommis KS, Butler AA. The importance of keeping time in the liver. Endocrinology (United States). 2021;162(2):1–10. https://doi.org/10.1210/endocr/bqaa230.
42. Tahara Y, Shibata S. Circadian rhythms of liver physiology and disease: experimental and clinical evidence. Nat Rev Gastroenterol Hepatol. 2016;13(4):217–26. https://doi.org/10.1038/nrgastro.2016.8.
43. Shi D, Chen J, Wang J, Yao J, Huang Y, Zhang G, Bao Z, Schroder EA. Circadian clock genes in the metabolism of non-alcoholic fatty liver disease. Front Physiol. 2019;10(May):423. https://doi.org/10.3389/fphys.2019.00423.
44. Mazzoccoli G, Vinciguerra M, Oben J, Tarquini R, Cosmo SD. Non-alcoholic fatty liver disease: the role of nuclear receptors and circadian rhythmicity. Liver Int. 2014;34(8):1133–52. https://doi.org/10.1111/liv.12534.
45. Roumagnac P, Weill FX, Dolecek C, Baker S, Brisse S, Chinh NT, Le TA, Acosta CJ, Farrar J, Dougan G, Achtman M. Evolutionary history of Salmonella typhi. Science. 2006;314(Nov):1304–9.
46. Pérez-Mendoza M, Rivera-Zavala JB, Díaz-Muñoz M. Daytime restricted feeding modifies the daily variations of liver gluconeogenesis: adaptations in biochemical and endocrine regulators. Chronobiol Int. 2014;31(7):815–28. https://doi.org/10.3109/07420528.2014.908898.
47. Pérez-Mendoza M, Rivera-Zavala JB, Rodríguez-Guadarrama AH, Montoya-Gomez LM, Carmona-Castro A, Díaz-Muñoz M, Miranda-Anaya M. Daily cycle in hepatic lipid metabolism in obese mice, neotomodon alstoni: sex differences. Chronobiol Int. 2018;35(5):643–57. https://doi.org/10.1080/07420528.2018.1424178.
48. Yuan Q, Pan A, Fu Y, Dai Y. Anatomy and physiology of the pancreas. In: Li M, Lu L, Xiao Y, Fu D, Zhang H, editors. Integrative pancreatic intervention therapy. Amsterdam: Elsevier; 2020. https://doi.org/10.1016/B978-0-12-819402-7.00001-2.
49. Mirzaei K, Xu M, Qi Q, De Jonge L, Bray GA, Sacks F, Qi L. Variants in glucose- and circadian rhythm-related genes affect the response of energy expenditure to weight-loss diets: the POUNDS LOST Trial1-3. Am J Clin Nutr. 2014;99(2):392–9. https://doi.org/10.3945/ajcn.113.072066.
50. Crosby P, Hamnett R, Putker M, Hoyle NP, Reed M, Karam CJ, Maywood ES, Stangherlin A, Chesham JE, Hayter EA, Rosenbrier-Ribeiro L, Newham P, Clevers H, Bechtold DA, O'Neill JS. Insulin/IGF-1 drives PERIOD synthesis to entrain circadian rhythms with feeding time. Cell. 2019;177(4):896–909.e20. https://doi.org/10.1016/j.cell.2019.02.017.
51. Ch R, Rey G, Ray S, Jha PK, Driscoll PC, Dos Santos MS, Malik DM, Lach R, Weljie AM, MacRae JI, Valekunja UK, Reddy AB. Rhythmic glucose metabolism regulates the redox circadian clockwork in human red blood cells. Nat Commun. 2021;12(1) https://doi.org/10.1038/s41467-020-20479-4.
52. Ch R, Chevallier O, Elliott CT. Metabolomics reveal circadian control of cellular metabolism. TrAC Trends Anal Chem. 2020;130:115986. https://doi.org/10.1016/j.trac.2020.115986.
53. Marcheva B, Ramsey KM, Buhr ED, Kobayashi Y, Su H, Ko CH, Ivanova G, Omura C, Mo S, Vitaterna MH, Lopez JP, Philipson LH, Bradfield CA, Crosby SD, Jebailey L, Wang X, Takahashi JS, Bass J. Disruption of the clock components CLOCK and BMAL1 leads to hypoinsulinaemia and diabetes. Nature. 2010;466(7306):627–31. https://doi.org/10.1038/nature09253.
54. Waddell H, Stevenson TJ, Mole DJ. The role of the circadian rhythms in critical illness with a focus on acute pancreatitis. Heliyon. 2023;9(4):e15335. https://doi.org/10.1016/j.heliyon.2023.e15335.
55. Kang R, Lotze MT, Zeh HJ, Billiar TR, Tang D. Cell death and DAMPs in acute pancreatitis. Mol Med. 2014;20(July–December):466–77. https://doi.org/10.2119/molmed.2014.00117.
56. Novosadová Z, Polidarová L, Sládek M, Sumová A. Alteration in glucose homeostasis and persistence of the pancreatic clock in aged MPer2 Luc mice. Sci Rep. 2018;8(1):1–11. https://doi.org/10.1038/s41598-018-30225-y.

57. Crnko S, Du Pré BC, Sluijter JPG, Van Laake LW. Circadian rhythms and the molecular clock in cardiovascular biology and disease. Nat Rev Cardiol. 2019;16(7):437–47. https://doi.org/10.1038/s41569-019-0167-4.
58. Škrlec I, Marić S, Včev A. Myocardial infarction and circadian rhythm. In: Visions cardiomyocyte—fundamental concepts of heart life and disease [working title]. 2019. https://doi.org/10.5772/intechopen.83393
59. Paul T, Lemmer B. Disturbance of circadian rhythms in analgosedated intensive care unit patients with and without craniocerebral injury. Chronobiol Int. 2007;24(1):45–61. https://doi.org/10.1080/07420520601142569.
60. Saxena R, Voight BF, Lyssenko V, Burtt NP, De Bakker PIW, Chen H, Roix JJ, Kathiresan S, Hirschhorn JN, Daly MJ, Hughes TE, Groop L, Altshuler D. Genome-wide association analysis identifies loci for type 2 diabetes and triglyceride levels. Science. 2007;316(5829):1331–6. https://doi.org/10.1126/science.1142358.
61. Alibhai FJ, Tsimakouridze EV, Chinnappareddy N, Wright DC, Billia F, O'Sullivan ML, Pyle WG, Sole MJ, Martino TA. Short-term disruption of diurnal rhythms after murine myocardial infarction adversely affects long-term myocardial structure and function. Circ Res. 2014;114(11):1713–22. https://doi.org/10.1161/CIRCRESAHA.114.302995.
62. Stujanna EN, Murakoshi N, Tajiri K, Xu D, Kimura T, Qin R, Feng D, Yonebayashi S, Ogura Y, Yamagami F, Sato A, Nogami A, Aonuma K. Rev-Erb agonist improves adverse cardiac remodeling and survival in myocardial infarction through an anti-inflammatory mechanism. PLoS One. 2017;12(12):1–19. https://doi.org/10.1371/journal.pone.0189330.

Sleep and Neuroinflammation

Arani Das, P. S. Mohan Raj, and Lal Chandra Vishwakarma

Abstract The intricate interplay between sleep regulation and neuroinflammation involves a symbiotic relationship, with immune-related molecules taking centre stage in this dynamic interaction. Beyond their conventional immune functions, cytokines and immune signalling components emerge as pivotal determinants of sleep patterns. Notably, interleukin-1 beta (IL-1β) and tumour necrosis factor-alpha (TNF-α), both pro-inflammatory cytokines, wield considerable influence over sleep modulation, encompassing key aspects such as slow-wave activity and non-rapid eye movement (NREM) sleep. A compelling facet of these cytokines is their adherence to a daily oscillation, hinting at intricate temporal entanglements with the sleep process. Recent inquiries into the NLRP3 inflammasome, recognized for its role in cytokine processing and secretion, unveil fresh perspectives on sleep regulation. By impacting the production of IL-1β, the NLRP3 inflammasome, activated by diverse cellular stressors, substantially shapes sleep architecture. Other inflammatory molecules such as adenosine, ATP, chemokines, neurotrophins, and reactive oxygen species give an insight into the complex interplay between sleep-modulating processes. Continued exploration of the complex interplay between these fundamental processes offers a more comprehensive understanding of their underlying mechanisms. This, in turn, opens possibilities for interventions to refine sleep quality and address associated sleep disorders.

Keywords Sleep · Neuroinflammation · Microglia · Astrocytes · IL-1β · TNF-α Adenosine

A. Das (✉) ·
Department of Physiology, All India Institute of Medical Sciences, Gorakhpur, Uttar Pradesh, India

P. S. Mohan Raj
Department of Biochemistry, All India Institute of Medical Sciences, Gorakhpur, Uttar Pradesh, India

L. C. Vishwakarma
Division of Sleep Medicine, Department of Medicine, University of Pennsylvania, Philadelphia, PA, USA

P. K. Jha, H. Bouaouda (eds.), *Circadian Rhythms, Sleep and Inflammation*, Progress in Inflammation Research 93,
https://doi.org/10.1007/978-3-031-74744-1_3

1 Introduction

Neuroinflammation refers to the inflammatory mechanisms that take place within the central nervous system (CNS) and involve both the innate and adaptive immune systems. Microglia, which are the brain's resident macrophages, and astrocytes, cells that typically have essential roles in various aspects of the brain, structure, metabolism, and physiology are the primary drivers of neuroinflammation. Sleep and inflammation share an intricate bidirectional relationship, with disruptions in one system impacting the other.

Activation of the immune system has a reciprocal relationship with sleep, and in turn, sleep influences both the innate and adaptive components of our body's defence system. When the immune system is stimulated by microbial challenges, it initiates an inflammatory response. The magnitude and timing of this response can lead to an increase in the duration and intensity of sleep, but it can also disrupt sleep patterns. The enhancement of sleep during an infection is believed to provide feedback to the immune system, promoting the host's defence mechanisms.

Immune-related molecules and inflammatory processes play key roles in regulating normal sleep–wake cycles and modulating electroencephalographic activity during sleep. Evidence shows pro-inflammatory cytokines promote sleep and slow-wave activity (SWA) response while anti-inflammatory cytokines attenuate these responses [1, 2]. Interleukin-1 beta (IL-1β) and tumour necrosis factor-alpha (TNF-α), two major pro-inflammatory cytokines, have been extensively studied for their influences on sleep regulation. Both molecules exhibit marked diurnal variations aligned with circadian rhythms and the sleep–wake cycle. Research indicates IL-1β promotes non-rapid eye movement (NREM) sleep while attenuating REM sleep [3, 4]. TNF-α similarly enhances slow-wave NREM sleep when administered exogenously [5]. Reciprocally, sleep deprivation leads to increased IL-1β and TNF-α expression in cortical and subcortical brain areas [3, 6]. This evidence highlights the key roles of inflammatory signalling molecules in physiological sleep–wake control. At the same time, irregular sleep–wake patterns and sleep disruption associated with chronic conditions like obesity and sleep apnea confer increased risk for inflammation-related morbidity [7, 8]. Ongoing research is focused on elucidating the intricate neuro-immune interactions governing sleep regulation and the repercussions of disturbed sleep on immune function. Future studies on these bidirectional relationships will uncover opportunities for more effective, immune-targeted therapies for sleep abnormalities and associated disorders.

In acute neuroinflammation, increased sleep response restores disrupted CNS homeostasis. Chronic neuroinflammation contributes to the development of neurodegenerative changes and various other disorders, often accompanied by sleep disturbances as a prominent comorbidity [9].

2 Glial Cells and Sleep

Sleep is a natural physiological state influenced by the actions of glial cells. Astrocytes and microglia, the two primary types of glial cells, both release inflammatory molecules and regulate sleep architecture [10]. The functions of astrocytes and microglia are also influenced by neuroinflammation, which in turn modulates sleep.

More than a century ago, Santiago Ramony Cajal, 1921 proposed the idea that astrocytes could influence sleep regulation by slowing down neural communication through their processes that extend into the synaptic cleft. Notably, modern studies, such as Pelluru et al.'s work utilizing optogenetics, have confirmed the role of astrocytes in modulating sleep [11]. Currently one widely used model for investigating the role of astrocytic gliotransmission in sleep regulation involves transgenic mice expressing the dominant-negative (dn)-SNARE domain in astrocytes. This model has been instrumental in research conducted by Halassa et al. [12] and Schmitt et al. [13].

Astrocytes are the primary source of extracellular ATP and adenosine, which contributes to the phasic activity in NREM sleep. Adenosine levels tend to rise during wakefulness and decrease during the sleep state. Another prominent role of astrocytes is removal of solutes and chemicals accumulated in the brain during the wake phase through the glymphatic system. Microglia, the resident immune cell of CNS, respond to pathological challenges, such as sleep loss, infection, neurodegenerative diseases, and traumatic injury [14]. Under physiological conditions, microglia are in a 'homeostatic' state, which is apparently a resting state, but their cytoplasmic processes are continuously exploring the CNS microenvironment. On pathological insults, such as trauma, sleep loss, and/or exposure to pathogens, microglia adopt a transition to an immunologically 'activated' state with increased proliferation, chemotactic migration, a phagocytic and secretory activity which includes the release of inflammatory mediators, reactive oxygen species, and/or trophic factors to promote brain immunity [15]. Current evidence shows microglia express circadian rhythms in circadian gene expression (Per1), pro-inflammatory cytokine expressions (IL-6, TNF-α, and IL-1β) which picks in the middle of the light/sleep phase. Microglia, in response to astrocyte-mediated ATP signalling, release cytokines including IL1-β and TNF-α to induce NREM sleep. These functions are mediated through purinergic P2X7 receptors on microglia [16] (Fig. 1). Other than microglia, IL1-β and TNF-α are also expressed in the hypothalamus and other brain areas to regulate autonomic, endocrine, and behavioural functions including sleep in health and disease [17].

Microglia and astrocytes display circadian variations in gene expression, morphology, and function that suggest active roles in sleep regulation. Studies have found rhythmic oscillations in microglial expression of clock genes like Per1 as well as the lysosomal protease Cathepsin-S, which peaked during the sleep phase in mice. Microglial process motility and phagocytic activity also follow a circadian pattern, with increased arborization during the active/dark phase in mice [18]. The

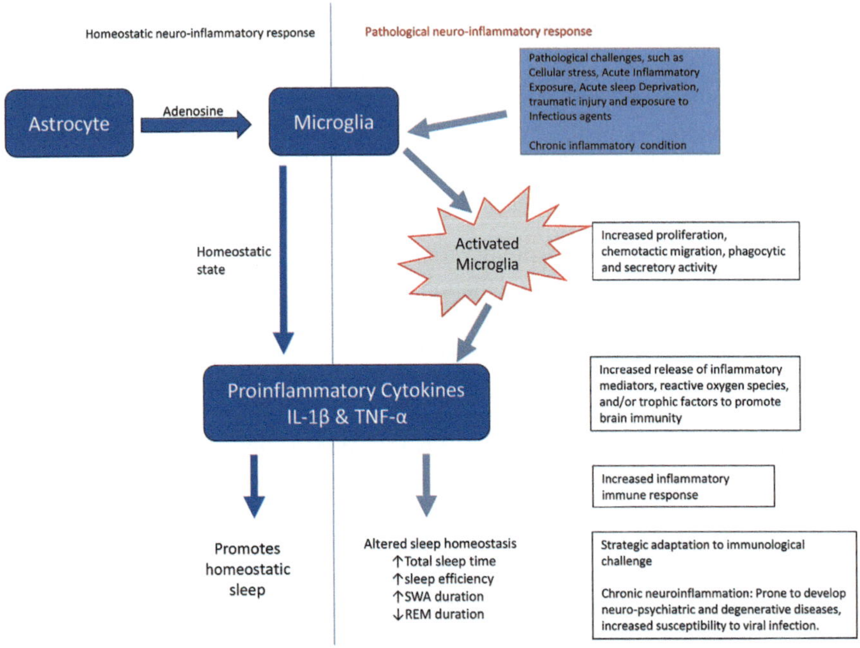

Fig. 1 Putative pathways for the effects of pathological challenge or stress on immune system involvement in sleep

normal nocturnal changes in microglia were prevented by blocking ATP signalling through the microglial P2Y12 receptor [18]. Astrocytes also show circadian patterns, with higher Per1 levels during sleep deprivation [19]. Several other contributions of microglia have been identified in impacting NREM sleep. Cathepsin-S, a member of the lysosomal cysteine protease family expressed by microglia, regulates dendritic spine plasticity and diffusion of chemicals in the brain during sleep. Microglia are involved in the phagocytic activity of opsonized synapses which mediates the downscaling of neurons during sleep [9].

The activation of microglia and astrocytes induced by sleep loss leads to the release of cytokines that can alter sleep homeostasis. For example, sleep deprivation was associated with microglial activation and increased release of pro-inflammatory cytokines like IL-1beta, IL-6, and TNF-alpha in rats [20]. The cytokine TNF-alpha has been shown to promote sleep in animal models while blocking TNF-alpha signalling reduces spontaneous sleep [21]. Astrocyte activation also contributes to sleep regulation through the release of ATP, which breaks down to adenosine, a neurotransmitter that builds up during wakefulness and promotes sleep [13]. Neuroinflammation and glial activation likely contribute to the pathophysiology of sleep disorders like obstructive sleep apnea. Levels of cytokines like IL-6 and TNF-alpha were elevated in patients with obstructive sleep apnea (OSA) compared to obese controls, independent of body weight, and reduced with continuous positive airway pressure (CPAP) treatment [22]. Children with OSA also showed increased systemic inflammation, which improved after adenotonsillectomy in those with resolved apnea [23]. This suggests a role for glial activation in the sleep fragmentation and hypoxia effects of OSA. Chronic sleep deprivation has been observed to cause glial-mediated synaptic remodelling in mice, with astrocytic and microglial phagocytosis of synaptic elements during sleep, which could impair cognition [24]. Persistent neuroinflammation after traumatic brain injury may also contribute to subsequent sleep disturbances through effects on orexin neurons and cytokine production [25]. Overall, complex bidirectional interactions exist between glial function, inflammatory mechanisms, sleep regulation, and neurological consequences of sleep loss.

3 Cytokines and Sleep

In 1909, a significant breakthrough marked the first indication of a central sleep-promoting substance. Researchers Ishimori in Japan and Piéron in France found that injecting cerebrospinal fluid (CSF) from sleep-deprived dogs into well-rested dogs induced sleep [1]. These initial findings were subsequently reinforced by several studies employing EEG recordings to objectively measure sleep in different animal species [2].

Simultaneously, scientists embarked on a quest to identify a sleep-promoting substance in the body's brain and fluids. In 1975, Pappenheimer successfully isolated a substance from the CSF and brain tissue of sleep-deprived animals, naming it 'factor S', with the 'S' designating its sleep-promoting properties [3]. Shortly thereafter, this factor was identified as a muramyl peptide, produced during the phagocytosis of macrophages and known for triggering the production of cytokines by macrophages [4, 5]. Over the subsequent years, researchers delved deeply into the study of cytokines, particularly IL-1 and TNF, in regulating homeostatic NREM sleep in animals.

To investigate the sleep-regulating properties of cytokines, researchers employed cytokine antagonists such as the IL-1 receptor antagonist (IL-1ra) or anti-IL-1 antibodies to suppress their biological actions. The blocking of the biological actions of cytokines like IL-1 and TNF resulted in a reduction in the quantity of physiological NREM sleep or a heightened NREM sleep rebound after a period of sleep deprivation and suppresses the REM sleep quantity. This underscores the critical role of cytokines in the regulation of sleep patterns [6]. Several other cytokines showed sleep-regulatory properties. Pro-inflammatory cytokines, such as IFN-, IL-2, IL-6, IL-15, and IL-18, tend to promote sleep and SWA while anti-inflammatory cytokines, including IL-4, IL-10, and IL-13 tend to attenuate sleep and SWA responses induced by sleep-promoting stimuli including sleep deprivation, pathogens, or pathogenic components [7, 8].

Animal and human research has indicated that pathogens or their components can influence sleep patterns. Observing heightened or intensified sleep response to acute infectious challenges has led to the theory that sleep is a beneficial mechanism for host defence and recovery. Findings by Toth et al. support this theory, as they discovered that rabbits inoculated with Escherichia coli bacteria experienced reduced morbidity and mortality when non-rapid eye movement (NREM) sleep was enhanced in response to the infection. One potential explanation for this supportive effect on the immune system is that sleep may facilitate allocating energy resources to the immune system. In contrast to the infection-induced increase in the duration and intensity of NREM sleep, various studies in animals and humans have consistently shown a simultaneous reduction in rapid eye movement (REM) sleep. This suppression of REM sleep may promote the development of fever or heat-generating mechanisms such as shivering which, in turn, enhances the immune response.

The two most studied pro-inflammatory cytokines, IL-1β and TNF-α are known to promote deeper NREM sleep.

3.1 IL-1β and Sleep

The IL-1 family of cytokines, identified as 11 members (IL-1a, IL-1b, IL-18, IL-33, IL-36a, IL-36b, IL-36 gamma (g), IL-1RA, IL-36RA, IL-37, IL-38), have function in immune responses, inflammation, and sleep homeostasis. Most IL-1 family members have pro-inflammatory properties, except that IL-37 and IL-1RA have anti-inflammatory actions. Evidence showed dysregulation of the IL-1 family in pathological states with disturbed sleep homeostasis including malignancies, brain damage, and cardiovascular diseases. Animal and human studies indicate that several IL-1 family members such as IL-1β, IL-1α, IL-18, and IL-37 are involved in regulating and modulating sleep architecture by their receptor-mediated downstream activity. Research over the past few decades indicates IL-1β also plays a key role in modulating physiological sleep [26].

Plasma levels of IL-1β demonstrate a clear circadian rhythm, with peak levels coinciding with the timing of increased sleep drive and sleep onset. Collective evidence suggests sleep deprivation increases the expressions of IL-1β in the cortex and other brain regions, peripheral tissue in most of the species including rabbits, cats, monkeys, mice, rats, and humans. Chronic sleep deprivation also increases IL-1β in the rodent cortex. Exogenous administration of infectious agent or gram-negative bacterial cell wall component lipopolysaccharide (LPS) results in increased expression of IL-1β in the cortex and hypothalamus of mice brains. LPS and influenza infection increase NREM sleep and alter slow-wave sleep architecture. These findings suggest a relationship between infection and sleep disturbances mediated by inflammatory immune responses.

Administration of exogenous IL-1β has been found to enhance non-rapid eye movement (NREM) sleep and increase electroencephalographic (EEG) slow-wave activity (a marker of deep NREM sleep) in rabbits, rats, mice, and other species [3, 4]. Microinjection of IL-1β directly into specific brain areas critical for sleep regulation, including the dorsal raphe nucleus, locus coeruleus, and preoptic area of the hypothalamus, induces NREM sleep in rats and guinea pigs. This evidence indicates IL-1β can directly modulate neuronal activity to promote NREM sleep. These findings are also associated with decreased rapid eye movement (REM) sleep, though the exact mechanism is still unknown. Blocking IL-1β signalling using an IL-1 receptor antagonist or neutralizing antibodies against IL-1β attenuates spontaneous NREM sleep in rabbits and rats. Mice lacking the IL-1 receptor type 1 exhibit reduced spontaneous NREM sleep compared to wild-type mice. Furthermore, sleep deprivation leads to increased IL-1β gene expression and protein levels in cortical, hypothalamic, brainstem, and other brain regions regulating sleep and wakefulness. Taken together, the evidence indicates IL-1β plays an important physiologic role in promoting and consolidating normal NREM sleep. Ongoing research is aimed at elucidating the precise molecular and neuronal mechanisms mediating these sleep-regulatory effects.

3.1.1 IL-1β in Sleep Disorders

In addition to its role in normal sleep regulation, evidence suggests dysregulated IL-1β signalling contributes to certain sleep disorders. Patients with obstructive sleep apnea (OSA) exhibit significantly elevated levels of IL-1β compared to controls. Plasma IL-1β levels correlate positively with OSA disease severity and subjective daytime sleepiness. Treatment of OSA with continuous positive airway pressure (CPAP) reduces IL-1β to normal levels in parallel with symptom improvement [22].

The sleep-enhancing effects of IL-1β are thought to be mediated in part through downstream signalling pathways involving IL-1 receptor activation, especially the IL-1 receptor type 1, as well as nuclear factor kappa B (NF-kB) signalling [26]. Dysfunction in these pathways could theoretically lead to hypersomnia and excessive daytime sleepiness as seen in OSA. Further research into the role of IL-1β signalling in sleep disorders could elucidate novel therapeutic targets for improved treatment.

3.2 *TNF-α and Sleep*

The cytokine tumour necrosis factor-alpha (TNF-α) is a well-established regulator of normal sleep and circadian biology, in addition to its functions in inflammation and immunity. TNF-α is produced by neurons, glia, and other cell types and demonstrates a clear circadian rhythm, with levels peaking during early nocturnal sleep among humans and rodents [1]. The altered expression of TNF-α is associated with dysregulated sleep and SWA in pathologies such as cancer, major depression, cardiovascular disease, and stroke. Studies in rat models showed higher TNF-a protein levels in the cortex, hippocampus, and hypothalamus during the beginning of the light period when sleep propensity is highest as well as in acute sleep deprivation or chronic sleep restriction. LPS application to the peritoneum and intranasal administration of influenza increases the expressions of TNF-α in sleep-regulating brain regions of mice models such as the cortex, NTS, and hypothalamus. All this experimental evidence demonstrates the role of TNF-α in homeostatic sleep and sleep responses to pathogens.

TNF-α and IL-1β share similarity in their effect of increasing NREM sleep at the expense of REM sleep. Studies on the TNF-receptor-1 KO mice model demonstrate lower spontaneous NREM sleep and reduced sleep response to pathogens and sleep deprivation. However, those mice showed normal response to IL-1β with increasing NERM sleep suggesting there are no overlapping in sleep-regulating mechanisms. Disrupted sleep–wake patterns are also seen in mice lacking enzymes involved in TNF-α processing, like the metalloprotease TACE/ADAM17 [26].

3.2.1 TNF-α in Sleep Disorders

Elevated TNF-a levels are also found to be significantly associated and correlated with the severity of sleep apnea in humans. Dysfunction of the TNF-α system likely contributes to sleep disorders associated with inflammation. Patients with obstructive sleep apnea (OSA) display two- to-threefold elevated TNF-α levels that correlate with excessive daytime sleepiness. Treating OSA with continuous positive airway pressure normalizes TNF-α levels independent of weight changes [2]. TNF-α is hypothesized to influence sleep by signalling through its receptor TNFR1, causing downstream nuclear factor kappa B (NF-kB) activation. Disruption of the TNF-α/TNFR pathway could promote sleep disturbances in OSA, insomnia, and other disorders exhibiting inflammatory dysregulation. Targeting TNF-α signalling may provide therapeutic opportunities for certain inflammation-related sleep pathologies [26].

4 Adenosine-ATP and Sleep

ATP has been hypothesized to play a prominent role in regulating sleep. Evidence suggests that prolonged wakefulness leads to higher extracellular levels of ATP and adenosine. ATP is primarily released from astrocyte-mediated inflammatory signalling into the extracellular space and modulates homeostatic sleep patterns. ATP acts through purine type 2X and type 2Y receptors [8]. Notably, the P2X7 receptor has been linked to the activation of the microglia and NLRP3 inflammasome and modulates NREM sleep homeostasis [16, 27].

Pharmacologically inhibiting the P2X7 receptor reduces sleep responses to sleep deprivation in rats, while its activation enhances spontaneous sleep and slow-wave activity (SWA). P2X7 receptor knockout mice exhibit reduced non-rapid eye movement (NREM) sleep and SWA responses to sleep deprivation [28]. CD39 is an enzyme responsible for converting ATP to adenosine diphosphate (ADP) and adenosine monophosphate (AMP). Another enzyme CD73 converts AMP to adenosine, which is the rate-limiting step. CD73 KO mice demonstrate reduced NREM sleep and SWA responses to sleep deprivation, suggesting that adenosine has a sleep-promoting role after prolonged wakefulness [26].

The sleep-regulating effect of adenosine has been well-described in multiple animal studies. Adenosine acts through A1 and A2a receptors to modulate NREM sleep. The extracellular adenosine promotes wakefulness through A1 receptors by decreasing cholinergic neuronal activity at basal forebrain neurons [29, 30]. However, studies on A1 receptor knockout mice showed normal NREM sleep time and slow-wave activity which provides evidence of redundancy in NREM sleep regulation [9]. Adenosine A2a receptors promote sleep by activating GABAergic neurons in the hypothalamus [31]. Studies on Adenosine A2a receptor knockout mice demonstrate the sleep-promoting effects of the A2a receptor [32].

5 NLRP3 Inflammasome and Sleep

Inflammasomes, a mediator of programmed cell death, are intracellular protein complexes with pattern recognition receptors to identify molecular pathogens. The nucleotide-binding domain and leucine-rich repeat protein-3 (NLRP3) inflammasome are most well-studied in their involvement in sleep regulation [33]. Several receptor-mediated pathways are identified in the process of NLRP3 activation, such as Toll-like receptor activation by LPS, interleukin 1 receptor 1 (IL-1R1) by IL-1β, or tumour necrosis factor receptor 2 (TNFR2) by TNF-α. A study on a knockout (KO) mice model showed, in response to LPS or infectious agents, NLRP3 inflammasome activation is a crucial mechanism for the downstream pathway which leads to IL-1β-mediated enhanced NREM sleep. Altogether recent evidence suggests that NLRP3 inflammasome provides a critical sensing mechanism that induces sleep and SWA in response to sleep deprivation and exposure to pathogens [34].

6 Chemokines and Sleep

This section explores the role of chemokines, specifically MCP-1 (Monocyte Chemoattractant Protein-1) and RANTES (Regulated on Activation, Normal T Cell Expressed and Secreted), in recruiting immune cells to sites of neuroinflammation which may impact sleep architecture and circadian rhythms.

Chemokines are important proteins supporting immune responses by guiding immune cells to regions of infection or inflammation. Acting as signalling molecules, they bind to specific receptors on immune cells and direct them where needed [35]. MCP-1 draws monocytes, a kind of white blood cell, to inflammatory sites. It's produced by numerous cell types, including macrophages, endothelial cells, and fibroblasts [36]. RANTES acts by recruiting T lymphocytes, eosinophils, and basophils to inflamed tissues. It's produced by various immunological and non-immune cells [36]. Studies have demonstrated chemokines, such as MCP-1 and RANTES, can alter sleep patterns through their participation in the immune response. Elevated chemokine levels linked to inflammation can disrupt normal sleep patterns, resulting in sleep disruptions and circadian rhythm abnormalities [37]. Neuroinflammation with raised chemokines is also linked with altered sleep patterns, producing difficulties falling asleep, interrupted sleep, or numerous overnight awakenings [38]. Chronic inflammation and increased chemokine levels can alter the immunological clock, compromising sleep timing and quality [36]. Understanding the delicate interplay between chemokines, inflammation, and sleep is vital for discovering the processes behind sleep disorders and changes in circadian rhythms. Further study is necessary to understand particular routes via which chemokines regulate sleep and circadian rhythms, offering options for new therapeutic therapies.

7 Neurotrophins and Sleep

Regulation of sleep involves a complicated web of neurobiological processes, and current research indicates that neurotrophins, such as nerve growth factor (NGF) and brain-derived neurotrophic factor (BDNF), may have substantial roles in both sleep control and the development of sleep disorders. Neurotrophins are a family of proteins necessary for the survival, growth, and function of neurons. Chronic inflammation, which can diminish BDNF levels, is an example of how neuroinflammation can negatively impair neurotrophic support [39]. Disruptions in neuroinflammation and neurotrophins levels can interfere with sleep regulation through interrelated processes. This connection is particularly relevant in neurological illnesses where sleep disruptions are widespread [36].

In the following section, we discuss the role of BDNF and NGF sleep regulation.

7.1 *BDNF and Sleep*

BDNF is a well-known neurotrophin supporting neuronal survival, growth, and synaptic plasticity. While continuing research is studying the direct association between BDNF and sleep, promising findings are suggesting its participation in sleep regulation. BDNF levels tend to increase during Rapid Eye Movement (REM) sleep, associated with vivid dreams and cognitive processing [40]. This implies a potential contribution to memory consolidation and cognitive functions connected to sleep. Studies indicate that greater BDNF levels may be connected to improved sleep quality and cognitive performance [41], suggesting a role in strengthening sleep-related cognitive functions. BDNF is related to sleep-dependent memory consolidation and synaptic plasticity, important for memory consolidation during sleep [38]. BDNF might modulate circadian rhythm, impacting sleep–wake cycles [42].

7.2 *NGF and Sleep*

NGF, less thoroughly studied compared to BDNF, has drawn attention in sleep research due to its function in the proliferation and survival of sensory and sympathetic neurons. While not directly linked to sleep regulation, NGF's impact on pain perception is notable. Dysregulated NGF levels may alter chronic pain issues, impacting sleep quality [43].

Emerging research reveals a relationship between BDNF dysregulation and depression, a condition commonly characterized by sleep problems. Dysregulated BDNF levels are implicated in mood disorders such as depression [44]. Conditions including Alzheimer's and Parkinson's disease, associated with altered neurotrophin levels, disrupt sleep by disrupting circadian rhythms and increasing daytime sleepiness [45]. Chronic pain associated with sleep difficulties may be altered by NGF, which plays a role in pain perception. For instance, fibromyalgia, characterized by widespread pain, is associated with sleep disruptions due to heightened pain perception [46]. Neurotrophins like BDNF and NGF are emerging as crucial players in understanding sleep regulation and the pathophysiology of sleep disorders. Although the specific mechanisms and therapeutic implications are still issues of ongoing research, examining the connection between neurotrophins and sleep holds great opportunities for expanding our understanding of sleep biology and developing tailored therapies for sleep-related diseases.

8 Neuroinflammatory Proteins and Sleep

Neuroinflammatory proteins, particularly Serum Amyloid A (SAA), have been attracting attention due to their participation in inflammation and the potential ramifications on sleep quality. This section goes into the involvement of neuroinflammatory proteins in inflammation and their influence on sleep.

Serum Amyloid A (SAA) is an acute-phase protein produced predominantly by the liver in response to inflammatory stimuli, like infection or tissue injury. While its major function is linked to the acute-phase response, evidence indicates that SAA has broad impacts on the immune system and inflammation. SAA functions as a pro-inflammatory cytokine, promoting the production of other inflammatory mediators such as interleukin-6 (IL-6) and tumour necrosis factor-alpha (TNF-α) [47], making it a critical role in the inflammatory response. SAA can stimulate multiple immune cells, including monocytes and macrophages, further increasing inflammation [48]. Elevated SAA levels in the bloodstream are suggestive of systemic inflammation, commonly detected in circumstances including chronic inflammatory disorders, cardiovascular disease, and infections [49]. Chronic inflammation with elevated SAA levels may contribute to sleep fragmentation, generating repeated awakenings during the night and affecting sleep architecture. This interruption adds to diminished sleep efficiency and overall poor sleep quality. Elevated SAA levels and other inflammatory markers have been seen in patients with insomnia [50].

9 Reactive Oxygen Species (ROS) and Sleep

This section explores the role of Reactive Oxygen Species (ROS) and their ability to generate oxidative stress and inflammation in the brain. ROS are chemically reactive oxygen-containing molecules, natural by-products of cellular metabolism, playing crucial roles in different physiological processes. Excessive ROS generation, frequently owing to causes like environmental pollutants, inflammation, or cellular malfunction, can contribute to oxidative stress. Oxidative stress comes from an imbalance between ROS production and the body's ability to counteract them with antioxidants. This imbalance can damage biological components, including proteins, lipids, and DNA. In the brain, oxidative stress can activate inflammatory responses [51]. Oxidative stress in the brain can disturb normal sleep architecture, generating fragmented sleep with frequent awakenings. Some research implies a potential relationship between oxidative stress and insomnia, where higher oxidative stress may contribute to difficulties getting or staying asleep. Paradoxically, sleep problems owing to oxidative stress can result in excessive daytime sleepiness, impairing cognitive function and overall well-being. Elevated oxidative stress levels are connected with reduced sleep duration. Chronic oxidative stress may lead to poor sleep, severely compromising health. Chronic oxidative stress is linked into several sleep disorders, including sleep apnea, where oxidative damage to airway tissues contributes to sleep-disordered breathing [51].

10 Summary

The article provides an in-depth look at the complex interplay between sleep regulation and neuroinflammation. It explains how immune-related signalling molecules like cytokines are pivotal to normal sleep patterns. In particular, the pro-inflammatory cytokines interleukin-1 beta (IL-1β) and tumour necrosis factor-alpha (TNF-α) strongly promote non-rapid eye movement (NREM) sleep and enhance slow-wave activity, which is indicative of deep sleep. IL-1β and TNF-α levels demonstrate clear circadian rhythms aligned with sleep drive. Microglia and astrocytes, as the resident immune cells of the brain, play major roles in modulating sleep by releasing these and other inflammatory factors like ATP and adenosine. The nucleotide-binding domain and leucine-rich repeat protein-3 (NLRP3) inflammasome are important for processing IL-1β production and regulating sleep architecture in response to diverse cellular stressors. Other inflammatory molecules such as adenosine, chemokines, neurotrophins, and reactive oxygen species can also impact sleep. The dysfunction of these inflammatory pathways likely contributes to sleep disorders such as obstructive sleep apnea, which represent possible future therapeutic targets. Ongoing research continues to uncover the nuanced neuro-immune interactions governing healthy sleep regulation and architecture. In summary, the article provides a thorough overview of the latest understanding on the intricate bidirectional relationships between sleep mechanisms and neuroinflammation.

Compliance with Ethical Standards The authors declare no conflict of interest regarding the publication of this book chapter.

This chapter dose not involve human or animal subjects or experimental procedures. Therefore, ethical approval was not applicable.

References

1. Lange T, Dimitrov S, Born J. Effects of sleep and circadian rhythm on the human immune system. Ann N Y Acad Sci. 2010;1193:48–59.
2. Vgontzas AN, Papanicolaou DA, Bixler EO, Lotsikas A, Zachman K, Kales A, et al. Circadian interleukin-6 secretion and quantity and depth of sleep. J Clin Endocrinol Metab. 1999;84(8):2603–7.
3. Interleukin 1-receptor antagonist blocks interleukin 1-induced sleep and fever—PubMed [Internet] [cited 3 Sept 2023]. Available from: https://pubmed.ncbi.nlm.nih.gov/1825458/
4. Comparative, behavioural and electrocortical effects of tumor necrosis factor-alpha and interleukin-1 microinjected into the locus coeruleus of rat—PubMed [Internet] [cited 3 Sept 2023]. Available from: https://pubmed.ncbi.nlm.nih.gov/9042390/
5. Kubota T, Li N, Guan Z, Brown RA, Krueger JM. Intrapreoptic microinjection of TNF-alpha enhances non-REM sleep in rats. Brain Res. 2002;932(1–2):37–44.
6. Chen Z, Gardi J, Kushikata T, Fang J, Krueger JM. Nuclear factor-kappaB-like activity increases in murine cerebral cortex after sleep deprivation. Am J Physiol. 1999;276(6):R1812–8.
7. Insomnia and its impact on physical and mental health—PMC [Internet] [cited 3 Sept 2023]. Available from: https://www.ncbi.nlm.nih.gov/pmc/articles/PMC3972485/
8. Oxidative stress in obstructive sleep apnea and intermittent hypoxia—revisited—the bad ugly and good: implications to the heart and brain—PubMed [Internet] [cited 3 Sept 2023]. Available from: https://pubmed.ncbi.nlm.nih.gov/25155182/
9. Garofalo S, Picard K, Limatola C, Nadjar A, Pascual O, Tremblay MÈ. Role of glia in the regulation of sleep in health and disease. Compr Physiol. 2020;10(2):687–712.
10. Frank MG. The role of glia in sleep regulation and function. Handb Exp Pharmacol. 2019;253:83–96.
11. Pelluru D, Konadhode RR, Bhat NR, Shiromani PJ. Optogenetic stimulation of astrocytes in the posterior hypothalamus increases sleep at night in C57BL/6J mice. Eur J Neurosci. 2016;43(10):1298–306.
12. Halassa MM, Florian C, Fellin T, Munoz JR, Lee SY, Abel T, et al. Astrocytic modulation of sleep homeostasis and cognitive consequences of sleep loss. Neuron. 2009;61(2):213–9.
13. Schmitt LI, Sims RE, Dale N, Haydon PG. Wakefulness affects synaptic and network activity by increasing extracellular astrocyte-derived adenosine. J Neurosci Off J Soc Neurosci. 2012;32(13):4417–25.

14. Deurveilher S, Golovin T, Hall S, Semba K. Microglia dynamics in sleep/wake states and in response to sleep loss. Neurochem Int. 2021;143:104944.
15. Salter MW, Stevens B. Microglia emerge as central players in brain disease. Nat Med. 2017;23(9):1018–27.
16. Ingiosi AM, Opp MR, Krueger JM. Sleep and immune function: glial contributions and consequences of aging. Curr Opin Neurobiol. 2013;23(5):806–11.
17. Breder CD, Tsujimoto M, Terano Y, Scott DW, Saper CB. Distribution and characterization of tumor necrosis factor-alpha-like immunoreactivity in the murine central nervous system. J Comp Neurol. 1993;337(4):543–67.
18. Hayashi Y. Diurnal spatial rearrangement of microglial processes through the rhythmic expression of P2Y12 receptors. J Neurol Disord. 2013;1:20.
19. Fornacon-Wood I, Faivre-Finn C, O'Connor JPB, Price GJ. Radiomics as a personalized medicine tool in lung cancer: Separating the hope from the hype. Lung Cancer Amst Neth. 2020;146:197–208.
20. Wadhwa M, Prabhakar A, Ray K, Roy K, Kumari P, Jha PK, et al. Inhibiting the microglia activation improves the spatial memory and adult neurogenesis in rat hippocampus during 48 h of sleep deprivation. J Neuroinflamm. 2017;14:222.
21. Involvement of cytokines in slow wave sleep—PubMed [Internet] [cited 18 Oct 2023]. Available from: https://pubmed.ncbi.nlm.nih.gov/21854954/
22. Yokoe T, Minoguchi K, Matsuo H, Oda N, Minoguchi H, Yoshino G, et al. Elevated levels of C-reactive protein and interleukin-6 in patients with obstructive sleep apnea syndrome are decreased by nasal continuous positive airway pressure. Circulation. 2003;107(8):1129–34.
23. Kheirandish-Gozal L, Gileles-Hillel A, Alonso-Álvarez ML, Peris E, Bhattacharjee R, Terán-Santos J, et al. Effects of adenotonsillectomy on plasma inflammatory biomarkers in obese children with obstructive sleep apnea: A community-based study. Int J Obes 2005. 2015;39(7):1094–100.
24. Sleep loss promotes astrocytic phagocytosis and microglial activation in mouse cerebral cortex—PubMed [Internet] [cited 18 Oct 2023]. Available from: https://pubmed.ncbi.nlm.nih.gov/28539349/
25. Viola-Saltzman M, Watson NF. Traumatic brain injury and sleep disorders. Neurol Clin. 2012;30(4):1299–312.
26. Zielinski MR, Gibbons AJ. Neuroinflammation, sleep, and circadian rhythms. Front Cell Infect Microbiol. 2022;12:853096.
27. Pelegrin P. P2X7 receptor and the NLRP3 inflammasome: Partners in crime. Biochem Pharmacol. 2021;187:114385.
28. ATP and the purine type 2 X7 receptor affect sleep—PMC [Internet] [cited 28 Sept 2023]. Available from: https://www.ncbi.nlm.nih.gov/pmc/articles/PMC2980381/
29. Porkka-Heiskanen T, Strecker RE, McCarley RW. Brain site-specificity of extracellular adenosine concentration changes during sleep deprivation and spontaneous sleep: an in vivo microdialysis study. Neuroscience. 2000;99(3):507–17.
30. Adenosine and sleep—PubMed [Internet] [cited 28 Sept 2023]. Available from: https://pubmed.ncbi.nlm.nih.gov/20190965/
31. Kumar S, Rai S, Hsieh KC, McGinty D, Alam MN, Szymusiak R. Adenosine A2A receptors regulate the activity of sleep regulatory GABAergic neurons in the preoptic hypothalamus. Am J Physiol Regul Integr Comp Physiol. 2013;305(1):R31–41.
32. Sleep regulation in adenosine A2A receptor-deficient mice—PubMed [Internet] [cited 28 Sept 2023]. Available from: https://pubmed.ncbi.nlm.nih.gov/14663019/
33. Zielinski MR, Gerashchenko D, Karpova SA, Konanki V, McCarley RW, Sutterwala FS, et al. The NLRP3 inflammasome modulates sleep and NREM sleep delta power induced by spontaneous wakefulness, sleep deprivation and lipopolysaccharide. Brain Behav Immun. 2017;62:137–50.

34. Zielinski MR, Gerashchenko D, Patel D, Torres K, Desrosiers G. 0219 mice lacking IL-18 have reduced sleep and slow-wave activity responses to sleep promoting stimuli. Sleep. 2019;42(Suppl_1):A90.
35. Soehnlein O, Lindbom L. Phagocyte partnership during the onset and resolution of inflammation. Nat Rev Immunol. 2010;10(6):427–39.
36. Ibarra-Coronado EG, Pantaleón-Martínez AM, Velazquéz-Moctezuma J, Prospéro-García O, Méndez-Díaz M, Pérez-Tapia M, et al. The bidirectional relationship between sleep and immunity against infections. J Immunol Res. 2015;2015:678164.
37. Opp MR, Krueger JM. Sleep and immunity: a growing field with clinical impact. Brain Behav Immun. 2015;47:1–3.
38. Krueger JM, Majde JA. Humoral links between sleep and the immune system: research issues. Ann N Y Acad Sci. 2003;992:9–20.
39. Chakraborti S, Chakraborti T, Mandal M, Mandal A, Das S, Ghosh S. Protective role of magnesium in cardiovascular diseases: a review. Mol Cell Biochem. 2002;238:163–79.
40. Abel T, Havekes R, Saletin JM, Walker MP. Sleep, plasticity and memory from molecules to whole-brain networks. Curr Biol CB. 2013;23(17):R774–88.
41. Viviani B, Gardoni F, Marinovich M. Cytokines and neuronal ion channels in health and disease. Int Rev Neurobiol. 2007;82:247–63.
42. Scheller J, Chalaris A, Schmidt-Arras D, Rose-John S. The pro- and anti-inflammatory properties of the cytokine interleukin-6. Biochim Biophys Acta. 2011;1813(5):878–88.
43. Understanding fibromyalgia and its related disorders. Prim Care Companion J Clin Psychiatry. 2008;10(2):133–44.
44. Duman RS, Aghajanian GK, Sanacora G, Krystal JH. Synaptic plasticity and depression: new insights from stress and rapid-acting antidepressants. Nat Med. 2016;22(3):238–49.
45. Scullin MK, Bliwise DL. Sleep, cognition, and normal aging: integrating a half century of multidisciplinary research. Perspect Psychol Sci J Assoc Psychol Sci. 2015;10(1):97–137.
46. Linsenmeyer TA. Use of botulinum toxin in individuals with neurogenic detrusor overactivity: State of the art review. J Spinal Cord Med. 2013;36(5):402–19.
47. Uhlar CM, Whitehead AS. Serum amyloid A, the major vertebrate acute-phase reactant. Eur J Biochem. 1999;265(2):501–23.
48. Connolly M, Mullan RH, McCormick J, Matthews C, Sullivan O, Kennedy A, et al. Acute-phase serum amyloid A regulates tumor necrosis factor α and matrix turnover and predicts disease progression in patients with inflammatory arthritis before and after biologic therapy. Arthritis Rheum. 2012;64(4):1035–45.
49. Meek RL, Urieli-Shoval S, Benditt EP. Expression of apolipoprotein serum amyloid A mRNA in human atherosclerotic lesions and cultured vascular cells: implications for serum amyloid A function. Proc Natl Acad Sci U S A. 1994;91(8):3186–90.
50. Irwin MR, Olmstead R, Carroll JE. Sleep disturbance, sleep duration, and inflammation: a systematic review and meta-analysis of cohort studies and experimental sleep deprivation. Biol Psychiatry. 2016;80(1):40–52.
51. Eisele HJ, Markart P, Schulz R. Obstructive sleep apnea, oxidative stress, and cardiovascular disease: evidence from human studies. Oxid Med Cell Longev. 2015;2015:608438.

Sleep and Circadian Rhythm in Peripheral Inflammation: Realm of Interconnected Frontiers

Atul Pandey, Ashish Kothari, and Santosh K. Kar

Abstract Sleep, a fundamental physiological process, intricately links with various facets of health, extending beyond mere rest and rejuvenation. This chapter deals with the intricate interplay among sleep, circadian rhythm, and peripheral inflammation, investigating deep into their complex interactions. Extensive research underlines sleep's pivotal role in regulating immune function and maintaining bodily equilibrium, with disruptions correlating to shifts in inflammatory markers and susceptibility to inflammatory diseases. Unraveling this intricate network necessitates a comprehensive understanding of the molecular mechanisms governing the processes, with cytokines, chemokines, and immune cells emerging as key players. Recent technological advancements shed light on the molecular underpinnings of sleep-induced immune modulation. Interruptions in sleep processes have been implicated in disturbing immune responses, thereby contributing to the development of inflammatory disorders. This chapter employs an interdisciplinary approach to dissect how sleep influences immune signaling and reciprocally, exploring its implications for chronic inflammatory conditions and potential therapeutic avenues. This further scrutinizes the impact of sleep duration, quality, and architecture on peripheral inflammation, as well as how inflammatory mediators reciprocally shape sleep patterns and processes. Understanding these intricate relationships holds promise for targeted interventions aimed at mitigating the repercussions of sleep disturbances on inflammation, and vice versa, thereby shaping the landscape of health and disease.

A. Pandey (✉)
Department of Entomology, University of Kentucky, Lexington, KY, USA

School of Biotechnology, Jawaharlal Nehru University, New Delhi, India
e-mail: atulkp@umich.edu

A. Kothari
Department of Microbiology, All India Institute of Medical Sciences, Rishikesh, Uttarakhand, India

S. K. Kar
School of Biotechnology, Jawaharlal Nehru University, New Delhi, India

P. K. Jha, H. Bouaouda (eds.), *Circadian Rhythms, Sleep and Inflammation*, Progress in Inflammation Research 93,
https://doi.org/10.1007/978-3-031-74744-1_4

Keywords Sleep · Circadian rhythm · Immunity · Inflammation · Inflammatory markers

1 Introduction

Sleep, a fundamental physiological process, is intricately connected to various aspects of health. Its significance extends far beyond the realms of rest and recovery, permeating into intricate physiological processes that govern overall health and well-being. Among the multifaceted dimensions of sleep, an emerging frontier has garnered increasing scientific interest—the intricate interplay between sleep, circadian rhythm, and immune function. The significance of sleep in regulating immune function and maintaining homeostasis has been emphasized by a growing body of recent literatures [1–4]. Conversely, disruptions in sleep patterns, whether acute or chronic, have been associated with alterations in inflammatory markers and heightened susceptibility to oxidative stress, cell death, and inflammatory diseases [5–10]. To disentangle the complex interplay between sleep and inflammation, it is crucial to explore the molecular mechanisms that govern these processes. The molecular signaling pathways, including those involving cytokines, chemokines, and cellular components of the immune system, play pivotal roles in mediating the cross-talk between sleep and peripheral inflammation [11–14]. Moreover, recent advances in technology, such as high-throughput omics approaches, have provided unprecedented insights into the molecular underpinnings of sleep-related immune modulation [15, 16].

The peripheral immune system, with its intricate network of cells and signaling molecules, plays a pivotal role in defending the body against pathogens and maintaining homeostasis. At the same time, disruptions in sleep processes have been implicated in dysregulating these finely tuned immune responses, potentially contributing to a spectrum of inflammatory disorders, e.g. rheumatoid arthritis, inflammatory bowel disease, psoriasis, asthma, allergic rhinitis, systemic lupus erythematosus, Type 2 diabetes, cardiovascular diseases, chronic obstructive pulmonary disease, chronic kidney disease, neurodegenerative disorders such as Alzheimer's disease, autoimmune disorders like multiple sclerosis, inflammatory skin conditions such as eczema, chronic inflammatory demyelinating polyneuropathy, and periodontal disease [11, 17–19]. By employing an interdisciplinary approach that integrates immunology, neurobiology, and sleep science, we aim to uncover the enigmatic interplay between sleep and peripheral inflammatory processes. The chapter delves into the intricate relationship between sleep and immune signaling, exploring how each influences the other. It examines how sleep patterns can impact immune function and, conversely, how immune processes can affect sleep architecture. Furthermore, the chapter explores the implications of sleep disorders on chronic inflammatory conditions and investigates potential therapeutic strategies that capitalize on the interplay between sleep and inflammation to promote overall health. Additionally, it discusses how inflammatory mediators, both systemic and local, reciprocally affect sleep patterns and contribute to sleep-related disorders. By unraveling these complex connections, the chapter highlights the potential for targeted interventions aimed at mitigating the effects of sleep disturbances on inflammatory processes and vice versa, offering promising avenues for improving health outcomes.

2 Sleep and Immune Function

The human sleep–wake cycle consists of various stages, each with distinct neurophysiological characteristics [20–22]. Studies have indicated that inadequate or disrupted sleep can influence the immune system, leading to alterations in peripheral inflammation [1, 3, 11, 23–25]. Understanding the nuance of sleep architecture is crucial for unraveling the intricate links between sleep, circadian rhythm, and immune function processes [15, 26–28]. Sleep architecture plays a crucial role in maintaining optimal immune function, with a growing body of scientific evidence highlighting the intricate interplay between sleep patterns and the immune system [11, 29, 30]. During the different stages of sleep, the body undergoes a complex interplay of neuroendocrine and immune processes, influencing immune cell trafficking, cytokine production, and overall immune responsiveness [31, 32]. Moreover, sleep disturbances, such as insufficient sleep duration or poor sleep quality, have been linked to alterations in immune function, potentially increasing susceptibility to infections and inflammatory disorders [16, 33]. Recent studies have highlighted the role of specific immune cells, such as T cells and natural killer cells, in the context of sleep-related immune modulation [11, 34, 35]. Additionally, the discovery of sleep-associated regulatory pathways, including the circadian rhythm and sleep–wake cycles, has further deepened our understanding of the intricate connections between sleep architecture and immune function [28, 36, 37]. Overall, these findings contribute to the growing body of knowledge supporting the importance of adequate and quality sleep for maintaining optimal immune health.

Some recent studies have elucidated the bidirectional relationship between sleep and immune function, shedding light on the specific mechanisms through which sleep architecture influences immune responses [26]. One notable study conducted by Irwin et al. demonstrated that sleep disturbances, such as insufficient sleep duration or poor sleep quality, are associated with alterations in immune cell activity, cytokine production, and inflammatory processes [2, 30, 32]. Furthermore, the work of Besedovsky et al. [31] has highlighted the role of different sleep stages, emphasizing that specific components of sleep architecture, such as slow-wave sleep and rapid eye movement (REM) sleep, contribute uniquely to immune regulation [31]. These findings underscore the importance of considering not only the total duration of sleep but also the qualitative aspects of sleep in understanding its impact on immune function. As research in this field continues to evolve, these insights may have significant implications for developing strategies to enhance immune health through targeted interventions in sleep architecture.

3 Circadian Rhythms and Inflammatory Mediators

Like sleep, the body's circadian rhythms play a pivotal role in orchestrating various physiological processes, including immune responses. Disruptions in circadian rhythms, often associated with irregular sleep patterns, can contribute to an imbalance in inflammatory mediators [38–40]. This section explores how disturbances in sleep–wake cycles may impact the regulation of cytokines, chemokines, and other inflammatory signaling molecules. Circadian rhythms, basically an intrinsic biological oscillation, play a pivotal role in regulating various physiological processes, including immune responses and inflammation [38, 41–44]. The intricate interplay between circadian processes and inflammatory mediators has gained significant attention in recent scientific research. Studies have elucidated the temporal regulation of pro-inflammatory cytokines such as interleukin-6 (IL-6) and tumor necrosis factor-alpha (TNF-α), by the circadian clock, highlighting the existence of circadian modulation in immune function [44–48]. Further emerging evidence suggests for communication between the circadian system and the immune system, where disruptions in circadian rhythms can impact inflammatory processes and vice versa. Notably, some recent research demonstrated that the circadian control of inflammatory responses occurs through the CLOCK-BMAL1 complex [49–51]. Furthermore, the studies emphasized the influence of circadian disruption on the NLRP3 inflammasome, implicating circadian regulation in the modulation of inflammatory pathways [52–55]. Collectively, these studies provide valuable insights into the complicated relationship between circadian rhythm and mediators of inflammatory processes, paving the way for a deeper understanding of how circadian dysregulation may contribute to inflammatory disorders.

4 Sleep Deprivation, Immunological Memory, and Inflammatory Markers

Sleep deprivation, whether acute or chronic, has been linked to alterations in systemic inflammatory markers. Research has identified changes in levels of C-reactive protein, IL-6, and TNF-α following periods of insufficient sleep [7, 10, 23, 56]. The mechanisms through which sleep deprivation influences these markers are discussed, providing insights into the potential health implications [11, 57]. Sleep deprivation has been extensively studied in relation to its impact on inflammatory markers, revealing a complex interplay between sleep patterns and the immune system. Numerous recent studies have underscored the association between insufficient sleep and heightened levels of pro-inflammatory cytokines such as IL-6 and TNF-α [16, 33, 58]. Additionally, investigations into the molecular pathways implicated in this phenomenon have identified NF-κB activation and increased expression of toll-like receptors as potential mechanisms linking sleep deprivation to inflammation [1, 59, 60]. Notably, a longitudinal study by Simpson et al. [61]

demonstrated a bidirectional relationship, with inflammation also contributing to disrupted sleep patterns [61]. Moreover, emerging evidence, as reported by Lange et al., suggests that the circadian regulation of inflammatory responses may be disrupted under conditions of sleep loss [28]. Thus, the convoluted connections between sleep and inflammatory markers underline the need for a comprehensive understanding of these processes, with implications for both therapeutic interventions and public health strategies.

Sleep deprivation also increases susceptibility to infections by dysregulating various immune parameters [1, 31]. Adequate sleep is essential for the proper functioning of both innate and adaptive immunity. During sleep, the body produces more cytokines like interleukin-12 that promote immune memory formation. Conversely, sleep loss leads to elevated levels of inflammatory cytokines like interleukin-6, contributing to chronic low-grade inflammation. Immunological memory, the ability of the immune system to remember and respond more effectively to previously encountered pathogens, is also influenced by sleep. Sleep enhances the formation and maintenance of immunological memory by facilitating the proliferation and differentiation of antigen-specific T and B cells. Lack of sleep can impair the generation of long-lasting immunological memory, reducing the efficacy of vaccines and increasing susceptibility to recurrent infections [26, 31].

5 Multidirectional Communication Between Sleep, Circadian Rhythm, and Peripheral Inflammation

The relationship between sleep and peripheral inflammation is not unidirectional. In addition to poor sleep influencing inflammation, inflammatory processes can reciprocally impact sleep patterns. This bidirectional communication is well explored, highlighting the complex interplay between immune activation and sleep disturbances [1, 31]. Sleep and circadian rhythms are intimately linked to physiological processes. Circadian rhythms are 24-h cycles that regulate various biological functions, including sleep–wake cycles. Some recent studies have shed light on this connection. For example, disruption of circadian rhythms can lead to sleep disorders and vice versa. Pandey et al. [62] found that rapid eye movement (REM) sleep deprivation in rats altered the expression of circadian clock genes in the brain and liver [62]. Conversely, circadian misalignment due to jet lag or shift work impairs sleep quality and quantity. A study found that "jet lag results from a misalignment between the internal circadian clock and the external environment as a result of rapid travel across multiple time zones...leading to sleep disturbances" [63]. Shift work also causes circadian disruption and sleep deficiency. A study paper notes, "Shift work-induced circadian misalignment not only affects sleep and health, it also determines when people can exercise, eat, socialize, and perform other daily activities" [64]. It further states that shift workers with circadian misalignment experience "sleep deficiency, sympathovagal and hormonal imbalance, inflammation, impaired glucose metabolism, and dysregulated cell cycles" [64].

Multidirectional communication between sleep, circadian rhythm, and peripheral inflammation constitutes a complex interplay critical for maintaining homeostasis within the human body. Sleep and circadian rhythm are intricately linked, influencing each other in a bidirectional manner. While recent studies have highlighted the relationship between sleep and circadian rhythm, with both phenomena contributing to altered immune responses and heightened peripheral inflammation [65]. Moreover, emerging research has elucidated the impact of peripheral inflammation on sleep architecture and circadian rhythm regulation, further emphasizing the intricate connections within this triad. A study by Irwin et al. [2] demonstrated that sleep disturbances can lead to increased pro-inflammatory cytokines, thereby exacerbating peripheral inflammation [2]. Conversely, the work of Lange et al. [28] highlighted how heightened peripheral inflammation, such as that seen in chronic inflammatory conditions, can disrupt sleep patterns and circadian rhythms [28]. Additional investigations by Besedovsky et al. [31] and Smith et al. [66] have provided insights into the molecular mechanisms mediating these interactions, implicating the involvement of cytokines and clock genes [31, 66]. The circadian clock, governed by molecular oscillators, orchestrates the timing of immune cell activity and cytokine release. Thus, disruptions to circadian rhythm, whether due to shift work or lifestyle factors, have been linked to heightened inflammation. Conversely, inflammation has been shown to influence the molecular clock machinery, as demonstrated in research (Fig. 1 [67, 68]). Recent findings also point to the involvement of the gut–brain axis in mediating the multidirectional communication between circadian rhythm, sleep, and peripheral inflammation. The recent works highlights how alterations in the gut microbiome can impact sleep and circadian processes, while disturbances in sleep and circadian rhythms can, in turn, affect gut health and immune function [69, 70]. Collectively, these findings underscore the multidirectional communication between sleep, circadian rhythm, and body inflammation, shedding light on the intricate regulatory network that maintains physiological balance.

6 Genomics and Transcriptomics of Sleep, Circadian Rhythm and Inflammation

Recent advancements in the field of "Genomics of sleep, circadian rhythm, and inflammation" have unveiled intricate molecular mechanisms underlying the interplay between these fundamental physiological processes. Studies have elucidated the genetic underpinnings of sleep regulation, circadian rhythm synchronization, and the immune response, shedding light on their interconnectedness and implications for health and disease.

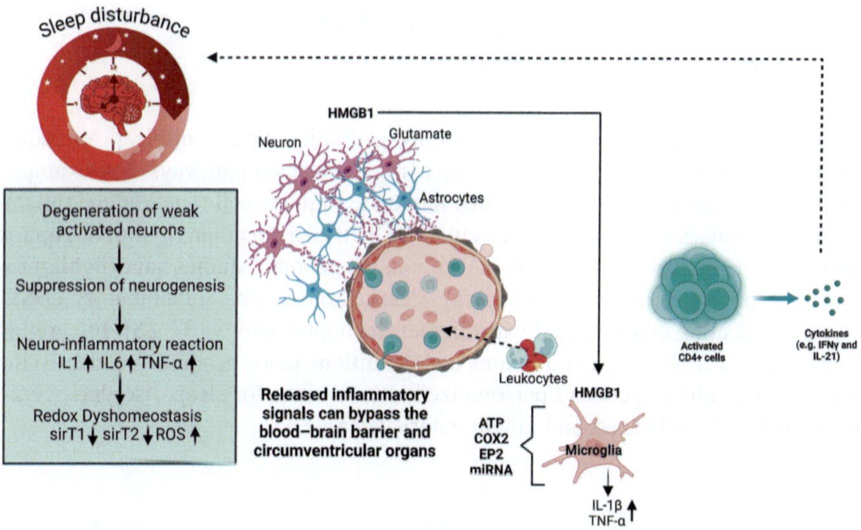

Fig. 1 Interconnected multidirectional relationship between sleep, circadian rhythm, and peripheral inflammation. *HMGB1* high mobility group box 1 protein, *IL-1* interleukin 1, *IL-6* interleukin 6, *IL-21* interleukin 21, *TNF-a* tumor necrosis factor-alpha, *SirT1* sirtuin 1, *SirT2* sirtuin 2, *ROS* reactive oxygen species, *ATP* adenosine triphosphate, *COX2* cyclooxygenase-2, *EP2* prostaglandin E2 receptor 2, *IL-1b* interleukin-1 beta, *CD4+* CD4 T lymphocytes, *IFN-g* interferon gamma and miRNA-micro RNA

6.1 Genomic Basis of Sleep and Circadian Rhythm Regulation

Genomic studies have identified genes associated with various aspects of sleep regulation, including sleep duration, quality, and architecture. For example, genome-wide association studies (GWAS) have linked variants in genes such as PER3 and NPAS2 to sleep duration and chronotype, highlighting the genetic basis of individual differences in sleep patterns [71, 72]. Furthermore, transcriptomic analyses have revealed dynamic changes in gene expression across sleep–wake cycles, implicating transcriptional regulation in sleep homeostasis [62, 73–76]. Advances in circadian rhythm genomics have elucidated the molecular machinery driving the endogenous circadian clock. Core clock genes, including CLOCK, BMAL1, PER, and CRY, orchestrate the rhythmic expression of thousands of genes involved in diverse physiological processes [77–80]. Recent studies have employed genomic techniques such as chromatin immunoprecipitation sequencing (ChIP-seq) and RNA sequencing (RNA-seq) to characterize the transcriptional regulatory networks governed by circadian clocks across different tissues and cell types [55, 81–83].

6.2 Integration of Sleep, Circadian Rhythm, and Inflammation Genomics

Recent efforts have focused on integrating genomics data from sleep, circadian rhythm, and inflammation to unravel their interconnected pathways. For example, studies have identified clock genes that modulate immune cell function and inflammatory responses, suggesting a molecular link between circadian rhythm disruption and immune dysregulation [84–86]. Additionally, genomic studies have highlighted shared genetic determinants underlying sleep disorders and inflammatory conditions, implicating common pathways in their pathophysiology [87, 88]. Integrating genomic data from different domains and conditions holds promise for identifying novel therapeutic targets and personalized interventions for sleep disorders, circadian rhythm disturbances, and inflammatory diseases.

6.3 Identification of Circadian Transcriptomes

Transcriptomic studies have unraveled the complex gene expression patterns governed by the circadian clock across various tissues and cell types. High-throughput sequencing techniques, such as RNA sequencing (RNA-seq), have facilitated the comprehensive characterization of circadian transcriptomes, revealing dynamic oscillations in gene expression over the course of the day [55, 78, 89]. While Zhang et al. 2014 [55] demonstrated the tissue-specific circadian gene expression profiles in the liver and brain, elucidating the temporal regulation of metabolic and neurobehavioral processes [55]. Further studies have found that a significant fraction, up to 68%, of protein-coding genes in the mammalian genome exhibit circadian oscillations in at least one tissue or condition or even when comparing transcriptomes from the same tissue under different perturbations, a major fraction of genes can produce circadian transcripts under some condition [89]. Recent technological advancements have also shed light on the molecular underpinnings of how the circadian clock regulates gene expression, including through rhythmic transcription factor binding, histone modifications, and chromosomal organization [78, 90].

6.4 Impact of Sleep Deprivation and Circadian Rhythm on Transcriptomic Signatures of Inflammatory Disorders

Sleep deprivation exerts profound effects on the transcriptome, disrupting normal gene expression patterns and influencing various physiological processes. Transcriptomic studies have elucidated the molecular mechanisms underlying the adverse effects of sleep loss on immune function and inflammatory signaling pathways. For instance, research by Cirelli et al. [91] identified transcriptional changes associated with sleep deprivation in brain regions involved in synaptic plasticity and neuroinflammation, providing insights into the neurobiological consequences of sleep disruption [91]. Transcriptomic profiling has facilitated the identification of molecular signatures associated with inflammatory disorders and their modulation by sleep and circadian rhythms. Integrative analyses combining transcriptomics with other omics data have revealed dysregulated pathways underlying inflammatory diseases, offering potential targets for therapeutic intervention. Noteworthy studies by Irwin et al. [16] and Patel et al. [92] elucidated transcriptomic alterations associated with chronic inflammatory conditions, highlighting the role of sleep and circadian dysfunction in disease pathogenesis [16, 92]. Thus, transcriptomic studies have significantly advanced our understanding of the complex interplay between sleep, circadian rhythm, and inflammation at the molecular level. By deciphering gene expression patterns and regulatory networks, these investigations provide valuable insights into the mechanisms linking sleep and circadian disturbances to inflammatory processes, offering novel avenues for targeted therapeutic interventions in inflammatory disorders.

7 Clinical Implications and Future Challenges

Understanding the intricate relationship between sleep-circadian processes and peripheral inflammation has important implications for clinical practice. This section discusses potential applications in the fields of sleep medicine, immunology, and chronic disease management. Furthermore, it outlines avenues for future research, emphasizing the need for longitudinal studies and intervention trials to elucidate causality and therapeutic potential. The investigation into the intricate interplay between sleep patterns, circadian rhythms, and peripheral inflammation holds significant clinical implications and offers avenues for future research. Numerous recent studies have highlighted the bidirectional relationship between sleep disturbances and inflammation, demonstrating how alterations in sleep duration and quality can influence immune responses and inflammatory processes. Additionally, disruptions in circadian rhythms, governed by the master clock in the suprachiasmatic nucleus, have been linked to variations in immune function and inflammatory markers [39, 49]. These findings have important implications for understanding and managing a range of health conditions, including autoimmune disorders, cardiovascular diseases, and neurodegenerative conditions [93–96]. Furthermore, emerging research suggests that interventions targeting sleep and circadian regulation may have therapeutic potential in mitigating inflammatory disorders [44, 97]. Future directions in this field involve exploring the precise molecular mechanisms that underlie the cross-talk between sleep, circadian rhythms, and inflammation [78, 98, 99]. Additionally, personalized approaches to managing sleep disorders and optimizing circadian health could be developed based on individual variations in immune response. The elucidation of these complex interactions holds promise for advancing our understanding of the pathophysiology of various diseases and informing targeted therapeutic interventions.

8 Conclusions

This chapter provides a comprehensive overview of the intricate interconnections between sleep, circadian rhythm, and peripheral body inflammation. Recognizing the bidirectional influences and exploring potential interventions can contribute to a holistic understanding of how sleep impacts inflammatory processes outside the central nervous system. As research in this field continues to evolve, it holds the promise of informing novel therapeutic approaches and preventive strategies for conditions influenced by both sleep and inflammation. The investigation into the intricate relationship between sleep, circadian rhythm, and peripheral body inflammation has unveiled a complex interplay within interconnected frontiers. Studies reviewed in this context have elucidated the potential pathways through which circadian rhythms modulate peripheral inflammatory responses, revealing the existence of intricate molecular and cellular signaling networks. Conversely, the impact of peripheral inflammation on sleep architecture suggests a reciprocal relationship, emphasizing the dynamic nature of these interactions. Furthermore, emerging evidence indicates that disturbances in these interconnected frontiers may contribute to the pathophysiology of various health conditions, including inflammatory disorders and sleep-related ailments.

While significant strides have been made in unraveling the complexities of the relationship between sleep, circadian rhythm, and peripheral body inflammation, numerous avenues for future research remain. Further exploration of the specific molecular mechanisms mediating these interactions, the identification of novel therapeutic targets, and the development of interventions to mitigate the adverse effects of dysregulation in these realms hold promise for advancing our understanding and improving clinical outcomes. The existing body of scientific literature supports the notion that disruptions in sleep patterns and circadian rhythms can significantly impact the regulation of inflammatory processes at the periphery. The bidirectional influence observed between these physiological phenomena underscores the need for a holistic understanding of the mechanisms governing immune responses and the temporal organization of sleep. As of now, we are at the crossroads of our understanding of sleep, circadian rhythm, and inflammation, which represents a multifaceted endeavor with implications for both basic science and clinical practice. Continued research in this field is essential for elucidating the underlying mechanisms, fostering the development of targeted interventions, and ultimately enhancing our ability to modulate these interconnected frontiers for improved health and well-being.

Compliance with Ethical Standards

Conflict of Interest The authors declare that they have no conflict of interest.

Ethical Approval This study does not involve any research involving human participants or animals performed by any of the authors.

References

1. Besedovsky L, Lange T, Born J. Sleep and immune function. Pflugers Arch. 2012;463:121–37. https://doi.org/10.1007/s00424-011-1044-0.
2. Irwin MR. Sleep and inflammation: partners in sickness and in health. Nat Rev Immunol. 2019;19:702–15. https://doi.org/10.1038/s41577-019-0190-z.
3. Opp MR, Baracchi F. Sleep and immune function. Curr Adv Sleep Biol. 2009; https://doi.org/10.1007/s00424-011-1044-0.
4. Zielinski MR, Krueger JM. Sleep and innate immunity. Front Biosci (Schol Ed). 2011;3:632–42. https://doi.org/10.2741/s176.
5. Haack M, Simpson N, Sethna N, Kaur S, Mullington J. Sleep deficiency and chronic pain: potential underlying mechanisms and clinical implications. Neuropsychopharmacology. 2020; https://doi.org/10.1038/s41386-019-0439-z.
6. Mullington JM, Simpson NS, Meier-Ewert HK, Haack M. Sleep loss and inflammation. Best Pract Res Clin Endocrinol Metab. 2010;24:775–84. https://doi.org/10.1016/j.beem.2010.08.014.
7. Pak VM, Paul S, Swieboda D, Balthazar MS, Wharton W. Sleep duration and biomarkers of inflammation in African American and white participants with a parental history of Alzheimer's disease. Alzheimers Dement (N Y). 2022;8:e12332. https://doi.org/10.1002/trc2.12332.
8. Pandey A, Kar SK. Rapid eye movement sleep deprivation of rat generates ROS in the hepatocytes and makes them more susceptible to oxidative stress. Sleep Sci. 2018; https://doi.org/10.5935/1984-0063.20180039.
9. Pandey A, Kumar D, Ray G, Kar SK. The flower pot method of REM sleep deprivation causes apoptotic cell death in the hepatocytes of rat. bioRxiv. 2019; https://doi.org/10.1101/375717.
10. Patel SR, Zhu X, Storfer-Isser A, Mehra R, Jenny NS, Tracy R, Redline S. Sleep duration and biomarkers of inflammation. Sleep. 2009;32:200–4. https://doi.org/10.1093/sleep/32.2.200.
11. Garbarino S, Lanteri P, Bragazzi NL, Magnavita N, Scoditti E. Role of sleep deprivation in immune-related disease risk and outcomes. Commun Biol. 2021;4:1–17. https://doi.org/10.1038/s42003-021-02825-4.
12. Krueger JM, Majde JA, Rector DM. Cytokines in immune function and sleep regulation. Handb Clin Neurol. 2011;98:229–40. https://doi.org/10.1016/B978-0-444-52006-7.00015-0.
13. Mackiewicz M, Naidoo N, Zimmerman JE, Pack AI. Molecular mechanisms of sleep and wakefulness. Ann N Y Acad Sci. 2008;1129:335–49. https://doi.org/10.1196/annals.1417.030.
14. Opp MR, Krueger JM. Sleep and immunity: A growing field with clinical impact. Brain Behav Immun. 2015; https://doi.org/10.1016/j.bbi.2015.03.011.
15. Haspel JA, Anafi R, Brown MK, Cermakian N, Depner C, Desplats P, Gelman AE, Haack M, Jelic S, Kim BS, Laposky AD, Lee YC, Mongodin E, Prather AA, Prendergast BJ, Reardon C, Shaw AC, Sengupta S, Szentirmai É, Thakkar M, Walker WE, Solt LA. Perfect timing: circadian rhythms, sleep, and immunity—an NIH workshop summary. JCI Insight. 2020;5:e131487. https://doi.org/10.1172/jci.insight.131487.

16. Irwin MR, Olmstead R, Carroll JE. Sleep disturbance, sleep duration, and inflammation: a systematic review and meta-analysis of cohort studies and experimental sleep deprivation. Biol Psychiatry. 2016;80:40–52. https://doi.org/10.1016/j.biopsych.2015.05.014.
17. Iranzo A. Sleep and neurological autoimmune diseases. Neuropsychopharmacology. 2020;45:129–40. https://doi.org/10.1038/s41386-019-0463-z.
18. Rheumatology TL. A wake-up call for sleep in rheumatic diseases. Lancet Rheumatol. 2022;4:e739. https://doi.org/10.1016/S2665-9913(22)00311-3.
19. Yin D, Chen S, Liu J. Sleep disturbances in autoimmune neurologic diseases: manifestation and pathophysiology. Front Neurosci. 2021;15:687536. https://doi.org/10.3389/fnins.2021.687536.
20. Brown RE, Basheer R, McKenna JT, Strecker RE, McCarley RW. Control of sleep and wakefulness. Physiol Rev. 2012;92:1087–187. https://doi.org/10.1152/physrev.00032.2011.
21. Patel AK, Reddy V, Shumway KR, Araujo JF. Physiology, Sleep stages. Treasure Island, FL: StatPearls; 2024.
22. Saper CB, Chou TC, Scammell TE. The sleep switch: hypothalamic control of sleep and wakefulness. Trends Neurosci. 2001;24:726–31. https://doi.org/10.1016/s0166-2236(00)02002-6.
23. Pandey AK, Kar SK. REM sleep deprivation of rats induces acute phase response in liver. Biochem Biophys Res Commun. 2011;410:242–6. https://doi.org/10.1016/j.bbrc.2011.05.123.
24. Prather AA, Leung CW. Association of insufficient sleep with respiratory infection among adults in the United States. JAMA Intern Med. 2016;176:850–2. https://doi.org/10.1001/jamainternmed.2016.0787.
25. Ragnoli B, Pochetti P, Pignatti P, Barbieri M, Mondini L, Ruggero L, Trotta L, Montuschi P, Malerba M. Sleep deprivation, immune suppression and SARS-CoV-2 infection. Int J Environ Res Public Health. 2022;19:904. https://doi.org/10.3390/ijerph19020904.
26. Ibarra-Coronado EG, Pantaleón-Martínez AM, Velazquéz-Moctezuma J, Prospéro-García O, Méndez-Díaz M, Pérez-Tapia M, Pavón L, Morales-Montor J. The bidirectional relationship between sleep and immunity against infections. J Immunol Res. 2015;2015:678164. https://doi.org/10.1155/2015/678164.
27. Imeri L, Opp MR. How (and why) the immune system makes us sleep. Nat Rev Neurosci. 2009;10:199–210. https://doi.org/10.1038/nrn2576.
28. Lange T, Dimitrov S, Born J. Effects of sleep and circadian rhythm on the human immune system. Ann N Y Acad Sci. 2010;1193:48–59. https://doi.org/10.1111/j.1749-6632.2009.05300.x.
29. Bryant PA, Trinder J, Curtis N. Sick and tired: Does sleep have a vital role in the immune system? Nat Rev Immunol. 2004;4:457–67. https://doi.org/10.1038/nri1369.
30. Irwin MR. Sleep disruption induces activation of inflammation and heightens risk for infectious disease: Role of impairments in thermoregulation and elevated ambient temperature. Temperature. 2023;10:198–234. https://doi.org/10.1080/23328940.2022.2109932.
31. Besedovsky L, Lange T, Haack M. The sleep-immune crosstalk in health and disease. Physiol Rev. 2019; https://doi.org/10.1152/physrev.00010.2018.
32. Irwin MR, Opp MR. Sleep health: reciprocal regulation of sleep and innate immunity. Neuropsychopharmacology. 2017;42:129–55. https://doi.org/10.1038/npp.2016.148.
33. Prather AA, Janicki-Deverts D, Hall MH, Cohen S. Behaviorally assessed sleep and susceptibility to the common cold. Sleep. 2015;38:1353–9. https://doi.org/10.5665/sleep.4968.
34. Dimitrov S, Lange T, Tieken S, Fehm HL, Born J. Sleep associated regulation of T helper 1/T helper 2 cytokine balance in humans. Brain Behav Immun (Special Issue on Sleep, Cytokines, and Immunity: Implications for Health). 2004;18:341–8. https://doi.org/10.1016/j.bbi.2003.08.004.
35. Tan H-L, Kheirandish-Gozal L, Gozal D. Sleep, sleep disorders, and immune function. In: Fishbein A, Sheldon SH, editors. Allergy and sleep: basic principles and clinical practice. Cham: Springer; 2019. p. 3–15. https://doi.org/10.1007/978-3-030-14738-9_1.
36. Khorasgani MR, Beikzadeh B. The impact of lifestyle on the immune system: focus on Islamic lifestyle: a narrative review. Int J Prev Med. 2023;14:105. https://doi.org/10.4103/ijpvm.ijpvm_11_22.

37. Malicki M, Karuga FF, Szmyd B, Sochal M, Gabryelska A. Obstructive sleep apnea, circadian clock disruption, and metabolic consequences. Metabolites. 2022;13:60. https://doi.org/10.3390/metabo13010060.
38. Fagiani F, Di Marino D, Romagnoli A, Travelli C, Voltan D, Di Cesare Mannelli L, Racchi M, Govoni S, Lanni C. Molecular regulations of circadian rhythm and implications for physiology and diseases. Sig Transduct Target Ther. 2022;7:1–20. https://doi.org/10.1038/s41392-022-00899-y.
39. Jerigova V, Zeman M, Okuliarova M. Circadian disruption and consequences on innate immunity and inflammatory response. Int J Mol Sci. 2022;23:13722. https://doi.org/10.3390/ijms232213722.
40. Wang D, Yin H, Wang X, Wang Z, Han M, He Q, Chen J, Xian H, Zhang B, Wei X, Yang B, Pan Y, Li J. Influence of sleep disruption on inflammatory bowel disease and changes in circadian rhythm genes. Heliyon. 2022;8:e11229. https://doi.org/10.1016/j.heliyon.2022.e11229.
41. Borrmann H, McKeating JA, Zhuang X. The circadian clock and viral infections. J Biol Rhythm. 2021;36:9–22. https://doi.org/10.1177/0748730420967768.
42. Hergenhan S, Holtkamp S, Scheiermann C. Molecular interactions between components of the circadian clock and the immune system. J Mol Biol. 2020;432:3700–13. https://doi.org/10.1016/j.jmb.2019.12.044.
43. Keller M, Mazuch J, Abraham U, Eom GD, Herzog ED, Volk H-D, Kramer A, Maier B. A circadian clock in macrophages controls inflammatory immune responses. Proc Natl Acad Sci USA. 2009;106:21407–12. https://doi.org/10.1073/pnas.0906361106.
44. Wang X-L, Li L. Circadian clock regulates inflammation and the development of neurodegeneration. Front Cell Infect Microbiol. 2021;11
45. Cutolo M. Circadian rhythms and rheumatoid arthritis. Joint Bone Spine. 2019;86:327–33. https://doi.org/10.1016/j.jbspin.2018.09.003.
46. Gibbs JE, Ray DW. The role of the circadian clock in rheumatoid arthritis. Arthritis Res Ther. 2013;15:205. https://doi.org/10.1186/ar4146.
47. Reinhardt ÉL, Fernandes PACM, Markus RP, Fischer FM. Night work effects on salivary cytokines TNF, IL-1β and IL-6. Chronobiol Int. 2019;36:11–26. https://doi.org/10.1080/07420528.2018.1515771.
48. Xing C, Zhou Y, Xu H, Ding M, Zhang Y, Zhang M, Hu M, Huang X, Song L. Sleep disturbance induces depressive behaviors and neuroinflammation by altering the circadian oscillations of clock genes in rats. Neurosci Res. 2021;171:124–32. https://doi.org/10.1016/j.neures.2021.03.006.
49. Curtis AM, Bellet MM, Sassone-Corsi P, O'Neill LAJ. Circadian clock proteins and immunity. Immunity. 2014;40:178–86. https://doi.org/10.1016/j.immuni.2014.02.002.
50. Kalsbeek A, la Fleur S, Fliers E. Circadian control of glucose metabolism. Mol Metab. 2014;3:372–83. https://doi.org/10.1016/j.molmet.2014.03.002.
51. Timmons GA, Carroll RG, O'Siorain JR, Cervantes-Silva MP, Fagan LE, Cox SL, Palsson-McDermott E, Finlay DK, Vincent EE, Jones N, Curtis AM. The circadian clock protein BMAL1 acts as a metabolic sensor in macrophages to control the production of pro IL-1β. Front Immunol. 2021;12
52. Katoku-Kikyo N, Lim S, Yuan C, Koroth J, Nakagawa Y, Bradley EW, Kikyo N. The circadian regulator PER1 promotes cell reprogramming by inhibiting inflammatory signaling from macrophages. PLoS Biol. 2023;21:e3002419. https://doi.org/10.1371/journal.pbio.3002419.
53. Nguyen KD, Fentress SJ, Qiu Y, Yun K, Cox JS, Chawla A. Circadian gene Bmal1 regulates diurnal oscillations of Ly6C(hi) inflammatory monocytes. Science. 2013;341:1483–8. https://doi.org/10.1126/science.1240636.
54. Pourcet B, Duez H. Circadian control of inflammasome pathways: implications for circadian medicine. Front Immunol. 2020;11:1630. https://doi.org/10.3389/fimmu.2020.01630.
55. Zhang R, Lahens NF, Ballance HI, Hughes ME, Hogenesch JB. A circadian gene expression atlas in mammals: implications for biology and medicine. Proc Natl Acad Sci USA. 2014;111:16219–24. https://doi.org/10.1073/pnas.1408886111.

56. Dolsen EA, Harvey AG. IL-6, sTNF-R2, and CRP in the context of sleep, circadian preference, and health in adolescents with eveningness chronotype: Cross-sectional and longitudinal treatment effects. Psychoneuroendocrinology. 2021;129:105241. https://doi.org/10.1016/j.psyneuen.2021.105241.
57. Dzierzewski JM, Donovan EK, Kay DB, Sannes TS, Bradbrook KE. Sleep inconsistency and markers of inflammation. Front Neurol. 2020;11:1042. https://doi.org/10.3389/fneur.2020.01042.
58. Haack M, Sanchez E, Mullington JM. Elevated inflammatory markers in response to prolonged sleep restriction are associated with increased pain experience in healthy volunteers. Sleep. 2007;30:1145–52.
59. Carvalho LA, Bergink V, Sumaski L, Wijkhuijs J, Hoogendijk WJ, Birkenhager TK, Drexhage HA. Inflammatory activation is associated with a reduced glucocorticoid receptor alpha/beta expression ratio in monocytes of inpatients with melancholic major depressive disorder. Transl Psychiatry. 2014;4:e344. https://doi.org/10.1038/tp.2013.118.
60. Irwin MR. Why sleep is important for health: a psychoneuroimmunology perspective. Annu Rev Psychol. 2015;66:143–72. https://doi.org/10.1146/annurev-psych-010213-115205.
61. Simpson NS, Banks S, Dinges DF. Sleep restriction is associated with increased morning plasma leptin concentrations, especially in women. Biol Res Nurs. 2010;12:47–53. https://doi.org/10.1177/1099800410366301.
62. Pandey A, Oliver R, Kar SK. Differential gene expression in brain and liver tissue of wistar rats after rapid eye movement sleep deprivation. Clocks Sleep. 2020; https://doi.org/10.3390/clockssleep2040033.
63. Baron KG, Reid KJ. Circadian misalignment and health. Int Rev Psychiatry. 2014;26:139–54. https://doi.org/10.3109/09540261.2014.911149.
64. James SM, Honn KA, Gaddameedhi S, Van Dongen HPA. Shift work: disrupted circadian rhythms and sleep—implications for health and well-being. Curr Sleep Med Rep. 2017;3:104–12. https://doi.org/10.1007/s40675-017-0071-6.
65. Shen Y, Lv Q, Xie W, Gong S, Zhuang S, Liu J, Mao C, Liu C. Circadian disruption and sleep disorders in neurodegeneration. Transl Neurodegener. 2023;12:8. https://doi.org/10.1186/s40035-023-00340-6.
66. Smith RP, Easson C, Lyle SM, Kapoor R, Donnelly CP, Davidson EJ, Parikh E, Lopez JV, Tartar JL. Gut microbiome diversity is associated with sleep physiology in humans. PLoS One. 2019;14:e0222394. https://doi.org/10.1371/journal.pone.0222394.
67. Fonken LK, Nelson RJ. The effects of light at night on circadian clocks and metabolism. Endocr Rev. 2014;35:648–70. https://doi.org/10.1210/er.2013-1051.
68. Gibbs JE, Blaikley J, Beesley S, Matthews L, Simpson KD, Boyce SH, Farrow SN, Else KJ, Singh D, Ray DW, Loudon ASI. The nuclear receptor REV-ERBα mediates circadian regulation of innate immunity through selective regulation of inflammatory cytokines. Proc Natl Acad Sci USA. 2012;109:582–7. https://doi.org/10.1073/pnas.1106750109.
69. Dinan TG, Cryan JF. Brain–gut–microbiota axis—mood, metabolism and behaviour. Nat Rev Gastroenterol Hepatol. 2017;14:69–70. https://doi.org/10.1038/nrgastro.2016.200.
70. Zheng D, Liwinski T, Elinav E. Interaction between microbiota and immunity in health and disease. Cell Res. 2020;30:492–506. https://doi.org/10.1038/s41422-020-0332-7.
71. Jones SE, Tyrrell J, Wood AR, Beaumont RN, Ruth KS, Tuke MA, Yaghootkar H, Hu Y, Teder-Laving M, Hayward C, Roenneberg T, Wilson JF, Del Greco F, Hicks AA, Shin C, Yun CH, Lee SK, Metspalu A, Byrne EM, Gehrman PR, Tiemeier H, Allebrandt KV, Freathy RM, Murray A, Hinds DA, Frayling TM, Weedon MN. Genome-wide association analyses in 128,266 individuals identifies new morningness and sleep duration loci. PLoS Genet. 2016; https://doi.org/10.1371/journal.pgen.1006125.
72. Lane JM, Jones SE, Dashti HS, Wood AR, Aragam KG, van Hees VT, Strand LB, Winsvold BS, Wang H, Bowden J, Song Y, Patel K, Anderson SG, Beaumont RN, Bechtold DA, Cade BE, Haas M, Kathiresan S, Little MA, Luik AI, Loudon AS, Purcell S, Richmond RC, Scheer FAJL, Schormair B, Tyrrell J, Winkelman JW, Winkelmann J, HUNT All in Sleep, Hveem K,

Zhao C, Nielsen JB, Willer CJ, Redline S, Spiegelhalder K, Kyle SD, Ray DW, Zwart J-A, Brumpton B, Frayling TM, Lawlor DA, Rutter MK, Weedon MN, Saxena R. Biological and clinical insights from genetics of insomnia symptoms. Nat Genet. 2019;51:387–93. https://doi.org/10.1038/s41588-019-0361-7.

73. Foo JC, Trautmann N, Sticht C, Treutlein J, Frank J, Streit F, Witt SH, De La Torre C, von Heydendorff SC, Sirignano L, Chen J, Müller-Myhsok B, Meyer-Lindenberg A, Witt CC, Gilles M, Deuschle M, Rietschel M. Longitudinal transcriptome-wide gene expression analysis of sleep deprivation treatment shows involvement of circadian genes and immune pathways. Transl Psychiatry. 2019;9:343. https://doi.org/10.1038/s41398-019-0671-7.
74. Hor CN, Yeung J, Jan M, Emmenegger Y, Hubbard J, Xenarios I, Naef F, Franken P. Sleep–wake-driven and circadian contributions to daily rhythms in gene expression and chromatin accessibility in the murine cortex. Proc Natl Acad Sci USA. 2019;116:25773–83. https://doi.org/10.1073/pnas.1910590116.
75. Jha PK, Valekunja UK, Ray S, Nollet M, Reddy AB. Single-cell transcriptomics and cell-specific proteomics reveals molecular signatures of sleep. Commun Biol. 2022;5:1–16. https://doi.org/10.1038/s42003-022-03800-3.
76. Maret S, Dorsaz S, Gurcel L, Pradervand S, Petit B, Pfister C, Hagenbuchle O, O'Hara BF, Franken P, Tafti M. Homer1a is a core brain molecular correlate of sleep loss. Proc Natl Acad Sci USA. 2007;104:20090–5. https://doi.org/10.1073/pnas.0710131104.
77. Pacheco-Bernal I, Becerril-Pérez F, Aguilar-Arnal L. Circadian rhythms in the three-dimensional genome: implications of chromatin interactions for cyclic transcription. Clin Epigenetics. 2019;11:79. https://doi.org/10.1186/s13148-019-0677-2.
78. Rijo-Ferreira F, Takahashi JS. Genomics of circadian rhythms in health and disease. Genome Med. 2019;11:82. https://doi.org/10.1186/s13073-019-0704-0.
79. Takahashi JS. Transcriptional architecture of the mammalian circadian clock. Nat Rev Genet. 2017;18:164–79. https://doi.org/10.1038/nrg.2016.150.
80. Vitaterna MH, Shimomura K, Jiang P. Genetics of circadian rhythms. Neurol Clin. 2019;37:487–504. https://doi.org/10.1016/j.ncl.2019.05.002.
81. Koike N, Yoo S-H, Huang H-C, Kumar V, Lee C, Kim T-K, Takahashi JS. Transcriptional architecture and chromatin landscape of the core circadian clock in mammals. Science. 2012;338:349–54. https://doi.org/10.1126/science.1226339.
82. Lu JY, Simon M, Zhao Y, Ablaeva J, Corson N, Choi Y, Yamada KYH, Schork NJ, Hood WR, Hill GE, Miller RA, Seluanov A, Gorbunova V. Comparative transcriptomics reveals circadian and pluripotency networks as two pillars of longevity regulation. Cell Metab. 2022;34:836–856.e5. https://doi.org/10.1016/j.cmet.2022.04.011.
83. Zhang Y, Chen G, Deng L, Gao B, Yang J, Ding C, Zhang Q, Ouyang W, Guo M, Wang W, Liu B, Zhang Q, Sung W-K, Yan J, Li G, Li X. Integrated 3D genome, epigenome and transcriptome analyses reveal transcriptional coordination of circadian rhythm in rice. Nucleic Acids Res. 2023;51:9001–18. https://doi.org/10.1093/nar/gkad658.
84. Fatima N, Sonkar GK, Singh S. Circadian mechanism disruption is associated with dysregulation of inflammatory and immune responses: a systematic review. Beni-Suef Univ J Basic Appl Sci. 2022;11:107. https://doi.org/10.1186/s43088-022-00290-4.
85. Inokawa H, Umemura Y, Shimba A, Kawakami E, Koike N, Tsuchiya Y, Ohashi M, Minami Y, Cui G, Asahi T, Ono R, Sasawaki Y, Konishi E, Yoo S-H, Chen Z, Teramukai S, Ikuta K, Yagita K. Chronic circadian misalignment accelerates immune senescence and abbreviates lifespan in mice. Sci Rep. 2020;10:2569. https://doi.org/10.1038/s41598-020-59541-y.
86. Scheiermann C, Kunisaki Y, Lucas D, Chow A, Jang J-E, Zhang D, Hashimoto D, Merad M, Frenette PS. Adrenergic nerves govern circadian leukocyte recruitment to tissues. Immunity. 2012;37:290–301. https://doi.org/10.1016/j.immuni.2012.05.021.
87. Ditmer M, Gabryelska A, Turkiewicz S, Białasiewicz P, Małecka-Wojciesko E, Sochal M. Sleep problems in chronic inflammatory diseases: prevalence, treatment, and new perspectives: a narrative review. J Clin Med. 2022;11:67. https://doi.org/10.3390/jcm11010067.

88. Jansen PR, Watanabe K, Stringer S, Skene N, Bryois J, Hammerschlag AR, de Leeuw CA, Benjamins JS, Muñoz-Manchado AB, Nagel M, Savage JE, Tiemeier H, White T, 23andMe Research Team, Tung JY, Hinds DA, Vacic V, Wang X, Sullivan PF, van der Sluis S, Polderman TJC, Smit AB, Hjerling-Leffler J, Van Someren EJW, Posthuma D. Genome-wide analysis of insomnia in 1,331,010 individuals identifies new risk loci and functional pathways. Nat Genet. 2019;51:394–403. https://doi.org/10.1038/s41588-018-0333-3.
89. Patel VR, Ceglia N, Zeller M, Eckel-Mahan K, Sassone-Corsi P, Baldi P. The pervasiveness and plasticity of circadian oscillations: the coupled circadian-oscillators framework. Bioinformatics. 2015;31:3181–8. https://doi.org/10.1093/bioinformatics/btv353.
90. Gagliano O, Luni C, Li Y, Angiolillo S, Qin W, Panariello F, Cacchiarelli D, Takahashi JS, Elvassore N. Synchronization between peripheral circadian clock and feeding-fasting cycles in microfluidic device sustains oscillatory pattern of transcriptome. Nat Commun. 2021;12:6185. https://doi.org/10.1038/s41467-021-26294-9.
91. Cirelli C, Tononi G. Differential expression of plasticity-related genes in waking and sleep and their regulation by the noradrenergic system. J Neurosci. 2000;20:9187–94. https://doi.org/10.1523/JNEUROSCI.20-24-09187.2000.
92. Patel NP, Grandner MA, Xie D, Branas CC, Gooneratne N. "Sleep disparity" in the population: poor sleep quality is strongly associated with poverty and ethnicity. BMC Public Health. 2010;10:475. https://doi.org/10.1186/1471-2458-10-475.
93. Lawther AJ, Phillips AJK, Chung N-C, Chang A, Ziegler AI, Debs S, Sloan EK, Walker AK. Disrupting circadian rhythms promotes cancer-induced inflammation in mice. Brain Behav Immun Health. 2022;21:100428. https://doi.org/10.1016/j.bbih.2022.100428.
94. Minakawa EN. Bidirectional relationship between sleep disturbances and Parkinson's disease. Front Neurol. 2022;13
95. Pan Y, Feng ZQ, Yuan Y, Hu GM, Jiang Y, Dong JC. Bidirectional relationship between circadian rhythm and frailty. NSS. 2023;15:949–53. https://doi.org/10.2147/NSS.S436488.
96. Walker WH, Walton JC, DeVries AC, Nelson RJ. Circadian rhythm disruption and mental health. Transl Psychiatry. 2020;10:1–13. https://doi.org/10.1038/s41398-020-0694-0.
97. Musiek ES, Ju Y-ES. Targeting sleep and circadian function in the prevention of Alzheimer disease. JAMA Neurol. 2022;79:835–6. https://doi.org/10.1001/jamaneurol.2022.1732.
98. Codoñer-Franch P, Gombert M, Martínez-Raga J, Cenit MC. Circadian disruption and mental health: the chronotherapeutic potential of microbiome-based and dietary strategies. Int J Mol Sci. 2023;24:7579. https://doi.org/10.3390/ijms24087579.
99. Zielinski MR, Gibbons AJ. Neuroinflammation, sleep, and circadian rhythms. Front Cell Infect Microbiol. 2022;12:853096. https://doi.org/10.3389/fcimb.2022.853096.

Neuro-inflammatory Response and Immune–Brain Crosstalk in Sleep Regulation

Pawan Kumar Jha and Hanan Bouâouda

Abstract Sleep and immune response are reciprocally linked. Lack of sleep impairs immune functions, increasing the organism's susceptibility to infection. The immune activation by microbial exposure triggers an inflammatory response, and based on its magnitude and time course, it modulates sleep. The inflammatory mediators (e.g., cytokines, chemokines) signal the brain to promote the sickness including fatigue and sleepiness. This is an adaptive response of the central nervous system (CNS) that enables organisms to facilitate recovery from infection/injury. In absence of microbial challenges, sleep appears to regulate the inflammatory mediators and maintain inflammatory homeostasis. This chapter gathers available data to explore the mutual interaction of sleep and immunity.

Keywords Circadian rhythms · Immunity · Infection · Inflammation · Sleep

1 Introduction

The evidence of the involvement of immune-related molecules in sleep regulation came from the experiments that identified the sleep-promoting factor "S" in cerebrospinal fluid (CSF) from sleep-deprived animals [1]. Later, this factor S was characterized as a bacterial cell wall peptidoglycan fragment known as muramyl peptide. Although muramyl peptide is not an endogenously produced sleep-promoting substance, the identification of muramyl peptide as a sleep-promoting factor linked sleep with immune functions. Subsequently, it was shown that muramyl peptide

P. K. Jha (✉)
Department of Systems Pharmacology & Translational Therapeutics, Perelman School of Medicine, University of Pennsylvania, Philadelphia, PA, USA

H. Bouâouda
Pharmacology Institute, Medical Faculty Heidelberg, Heidelberg University, Heidelberg, Germany

P. K. Jha, H. Bouaouda (eds.), *Circadian Rhythms, Sleep and Inflammation*, Progress in Inflammation Research 93,
https://doi.org/10.1007/978-3-031-74744-1_5

induces the synthesis and secretion of the pro-inflammatory cytokine interleukin 1 (IL-1) [2, 3]. Further studies demonstrated that central administration of IL-1 enhances non-rapid eye movement (NREM) sleep [4–6]. These findings indicate that sleep could be crucial to immune responsiveness and function.

As more becomes known about the sleep response to the infectious challenge and attenuation of the immune response following sleep disturbance [7–9], it is conceivable that immune and inflammatory molecules could be the regulators of the sleep–wake cycle and sleep participates in maintaining immune function. Herein, we describe the immune–brain interactions in sleep response to immune challenges and the effect of sleep loss on indices of inflammatory system activation.

2 Circadian Rhythms and Sleep–Wake Regulation

Most living organisms exhibit daily oscillations in their behavior, metabolism, and physiological functions, with a period close to 24 h even without external environmental cues (*Zeitgebers*) [10, 11]. In mammals, these daily oscillations are coordinated by a circadian system comprising a master clock located in the suprachiasmatic nuclei (SCN) of the hypothalamus, synchronizing input and output pathways. The master clock conveys its circadian signals to various other oscillators to impose daily changes in the temporal organization of behavioral, metabolism, and physiological processes [12–15]. Light is the major source of entrainment for circadian rhythms, although several other non-photic factors like food, temperature, and exercise have similar abilities [13, 16]. The retinal ganglion cells perceive the light and transduce this photic signal to the SCN through the retinohypothalamic tract [17, 18]. Through this anatomical path, the circadian rhythms generated within SCN neurons are entrained into an overt 24-h rhythm and these coordinated outputs are conveyed to the rest of the body via behavioral, neuroendocrine, and autonomic pathways [14, 19]. In this manner, SCN controls various behavioral and physiological circadian rhythms that include the sleep–wake cycle, glucose metabolism, and hormone secretion [12, 20, 21], cognitive processes such as mood [22], and learning and memory processing [23–25]. Overall, the circadian system enables living organisms to anticipate daily recurring events and adapt their behavioral and physiological functions accordingly.

Circadian rhythms exist in every cell of the body. At the molecular level, the circadian rhythms generated within an individual cell are encoded by the rhythmic expression of core clock genes and their autoregulatory positive and negative transcription/translation feedback loops (Fig. 1) [26–28]. The molecular mechanism of circadian rhythms within a cell is explained by the interaction of positive transcriptional regulators comprising Circadian Locomotor Output Cycle Kaput (CLOCK) and Brain Muscle Aryl Hydrocarbon Receptor Nuclear Translocator-Like1 (BMAL1, encoded by *Arntl* gene), and negative regulators such as PERIODs (PERs: PER1,2, and 3), CRYPTOCHROMEs (CRYs: CRY1 and 2), and REV-ERBs [19, 29, 30]. BMAL1 binds with CLOCK, forms heterodimers, and initiates the

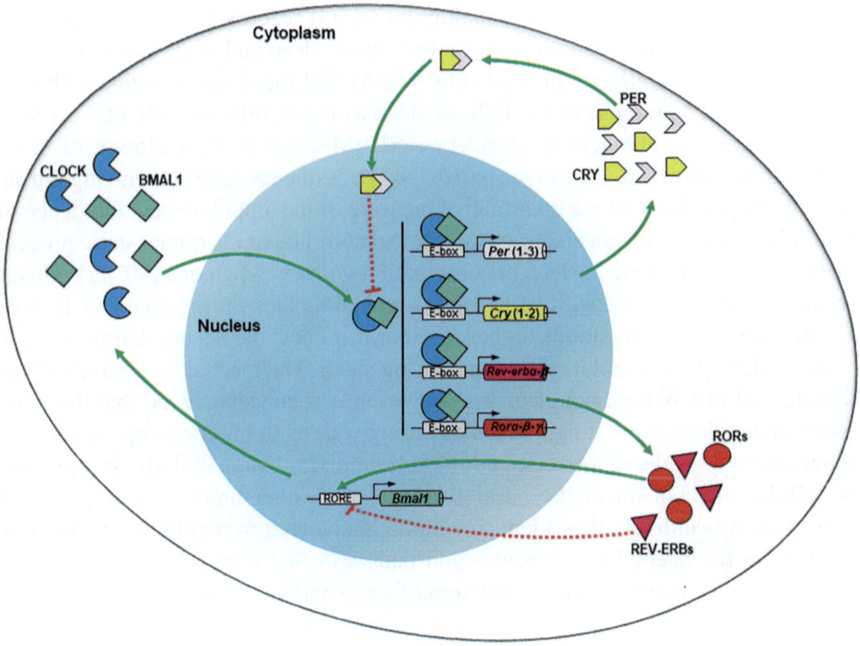

Fig. 1 The mammalian transcriptional-translational feedback loop (TTFL) of the circadian clock. In mammalian TTFL, CLOCK and BMAL1 form heterodimers and bind to the E-boxes in Per1–3 and Cry1–2 promoters, resulting in their transcriptions. PER and CRY proteins accumulate in the cytoplasm, form heterodimers, enter the nucleus, and inhibit their own transcription by suppressing CLOCK-BMAL1 activity. Additionally, CLOCK and BMAL1 also promote RORs and REV-ERBs transcription, which in turn activate and inhibit BMAL1 transcription, respectively. *BMAL1* brain and muscle ARNT-like 1; *CLOCK* circadian locomotor output cycles kaput; *PER* period; *CRY* cryptochrome; *REV-ERBs* reverse viral erythroblastosis oncogene products; *RORs* retinoic acid-related orphan receptors. Adapted from refs. [19, 29]

rhythmic transcription of *Per* and *Cry* by binding to the enhancer E-Box sequences located in the promotor regions of these genes [31, 32]. Subsequently, PERs and CRYs proteins accumulate in the cytoplasm, where they are phosphorylated by casein kinase Iε, form heterodimers, and translocate back to the nucleus to inhibit their own transcription, by acting on CLOCK:BMAL1 complex [27, 33, 34]. Another negative feedback loop involves the retinoic acid-related orphan nuclear receptors (*Rorα, β,* and *γ*) and Reverse Viral Erythroblastosis oncogene products (*Rev-erbα* and *β*), which are also direct transcriptional targets of CLOCK/BMAL1 heterodimers [35, 36]. The resultant RORs and REV-ERBs proteins regulate the transcription of *Bmal1* by binding to ROR response elements (ROREs) present in the *Bmal1* promoter [37]. Hence, RORs trigger the transcription of *Bmal1* [37, 38], whereas REV-ERBs repress this transcriptional process [36, 37].

The sleep–wake cycle is an example of circadian rhythms. Besides the circadian component, sleep also has a strong regulatory homeostatic component which can be understood by the observation that loss of sleep is compensated by extending subsequent sleep. The electroencephalogram (EEG) records the fluctuation of the brain's electrical activity and is used to distinguish sleep and wake states. In mammals, sleep is generally subdivided into NREM and rapid eye movement (REM) sleep. The NREM represents the bulk of sleep as it constitutes nearly 80% of total sleep. The slow wave activity (SWA) of NREM sleep is the hallmark of sleep homeostasis. This is EEG delta power (0.5–4 Hz frequency range) occurring during NREM sleep and provides a quantitative measure of the amplitude and incidence of SWA [39]. The SWA is an electrophysiological correlate of the homeostatic process that determines the time to sleep in combination with circadian one [40, 41]. This is because SWA increases with wakefulness and dissipates during sleep. It is even higher after sleep deprivation, suggesting that it reflects the accumulation of sleep pressure during wake and its discharge during sleep. The mechanistic details of the rise and fall of SWA are not clear yet. However, it is conceptualized that the brain accumulates sleep-promoting molecules during waking to initiate sleep. At present, adenosine is considered a compound that accumulates extracellularly as a product of cellular metabolism in the basal forebrain and other brain regions [42, 43]. Moreover, it is unlikely that adenosine alone is involved in regulating the homeostatic drive for sleep. The molecules and pathways that form part of the immune system are also involved in the regulation of sleep and vice versa.

3 Inflammation and Immunity

Inflammation is an essential immune response that enables survival during infection or other cellular insults and maintains tissue homeostasis by eliminating the initial cause of cell injury. The immune cells use a variety of immune molecules such as cytokines, chemokines, and other chemical mediators that modulate the complex immune response. The immune cells are broadly divided into two categories: (1) cells that form the unspecific, natural, or innate immune system include granulocytes (neutrophils, eosinophils, or basophils that develop into tissue mast cells), blood monocytes that develop into tissue macrophages (Langerhans cells in the skin, Kupffer cells in liver, microglia-like macrophages in the brain), dendritic cells and natural killer (NK) cells; and (2) cells that form the specific, acquired, adaptive immune system recognize a specific antigenic peptide by expressing unique receptors on their cell surface. The B lymphocytes or B cells develop in the bone marrow and express the B cell receptor; T lymphocytes or T cells develop in the thymus and express the T cell receptor. Moreover, specific non-immune cells like mucosal and skin epithelia produce antimicrobial peptides and complement factors to cope with substantial microbial exposure [44].

The tissue-resident macrophages sense the pathogenic invasion or stressful stimulus by expressing pattern recognition receptors (PRRs). The PRRs recognize the pathogen-associated molecular patterns (PAMPs) from microbes like bacteria or viruses or endogenous danger/damage-associated molecular patterns (DAMPs), which are released by stressed or injured cells [45]. This led to the activation of inflammatory signaling pathways like inflammasomes or nuclear factor-kappa B (NF-kB) followed by the release of cytokines, interferons with anti-viral activity, vasoactive mediators like prostaglandins (PGs), and chemokines that attract further leukocytes into inflammatory sites. Consequently, clinical signs of inflammation (redness, swelling, warming, pain, impaired function) appear, and phagocytosis or the release of antimicrobial substances takes place. The phagocytic and chemotactic responses of neutrophils and NK cells build the line of defense of innate immunity. The systemic innate response prompts the central nervous system to induce fever or sickness behaviors.

The adaptive immune system provides targeted and specific responses. The antigen-presenting cells (APCs) present fragments of pathogen and the major histocompatibility complex II (MHCII) to naive T cells. This leads to the differentiation of naive T cells (CD4 T cells) into effector T-helper cells (CD4 T-helper cells) which help the cytotoxic T cells (CD8 T cells), macrophages, and B cells to eliminate the pathogen. The activated B cells turn into plasma cells and produce antibodies to neutralize the soluble antigen. These antigen-specific memory T and B cells and antigen-specific antibodies constitute the basis of immunological memory. Based on the pattern of PAMPs and DAMPs, APCs release certain cytokines, thereby T cells and eventually the class of the antibodies, indicating the proximal activity of innate and adaptive immune system [46–48].

The inflammatory response triggered by local infection or tissue damage involving PAMPs and DAMPs is primarily protective and acts as a function of restoring physiological homeostasis [49, 50]. The cytokines, chemokines, vasoactive amines, and PGs are the inflammatory mediators that induce a variety of biological actions, e.g. recruitment of neutrophils from the circulation to affected tissues, and vascular changes to maintain homeostasis [51]. Some of these acute responses can be detected in circulation in the form of inflammatory markers and may act systemically to induce sickness behaviors. Like in the periphery, acute neuroinflammation is also considered protective. The microglia, resident macrophages, and astrocytes primarily command neuroinflammation and play a central role in many aspects of brain metabolism [52].

Persistent inflammation leads to loss of function of restoring homeostasis. The elevations of inflammatory markers have been reported in pathological conditions like type 2 diabetes, cardiovascular diseases, obesity, asthma, some chronic pain conditions, some forms of cancer, and neurodegenerative diseases [53]. The elevation of inflammatory markers is also reported in chronic sleep-deprived conditions [54–56]. In addition, the aging process also triggers inflammatory responses, a phenomenon termed inflammaging [57, 58]. The mechanism that drives these low grades of inflammation is not well known. It is argued that these responses are not mediated by classical inducers of inflammation like infectious agents or tissue injury, but rather cellular stress or its malfunctioning produces the DAMPs (heat-shock proteins, uric acid, amyloid beta, etc.) that activate PRRs [59, 60]. Moreover, metabolites (glucose and fatty acids) or byproducts of metabolites (cholesterol, ceramide, and uric acid) are also acting as DAMPs that may contribute to inflammation and this may explain the inflammation under metabolic abnormalities [61].

4 Sleep Loss and Immune Function

Several experimental studies have measured daily variations of immune parameters that provide strong links between circadian rhythms and the immune system [62, 63]. Perturbation of immune parameters like leucocytes in blood and various tissues, circulating cytokine levels, cytokine production in specific leukocytes, the concentration of antibodies, and complement factors are vastly investigated in both human and animal models of sleep manipulation. Although the methodology used for sleep manipulation vary widely, they are generally aimed at reducing sleep duration. Some studies induce total sleep deprivation, while others employ partial sleep deprivation, such as sleep restriction for a few hours or selective REM sleep deprivation. Broadly these divisions can be categorized into acute sleep deprivation/restriction if intervention is of 1 day or less and chronic sleep deprivation/restriction if protocols use more prolonged sleep manipulation. In this section, we gather information on sleep manipulation and immune responses.

4.1 Leukocyte Distribution

The number of leukocytes in the blood is widely studied for circadian rhythms and sleep studies. These cells display strong migratory capacity and travel in and out to various tissues and organs under different circumstances using the blood and the lymphatic system as travel routes. This migration and migration-driven acute changes in the blood count of leukocyte subsets display robust circadian components [63, 64]. Sleep influences the number of leucocytes in blood; this impact could be significant and consistent. Several studies in humans have reported that sleep disruption increases leucocyte counts [65–69]. Other studies reported no changes following sleep manipulation [70–74]. Contrary to reports from human subjects, the long duration of REM sleep deprivation in murine showed a dip in total cellularity in the blood [75]. There is the possibility that a longer duration of sleep deprivation may affect blood cell proliferation in rodents.

Among the subsets of leukocytes, monocytes, and lymphocytes levels either increased or remained unchanged following acute or chronic sleep deprivation in human subjects [76]. As for granulocyte subsets, basophils and eosinophils counts in the blood remain unchanged following sleep manipulation; however, many studies reported a rise in neutrophils following sleep deprivation [68, 69, 74, 77]. Again, some studies reported unchanged neutrophil levels [65, 71, 73, 78]. This discrepancy could be due to differences in methodologies or conditions.

The change in various leukocyte subsets in the blood during acute sleep manipulation is indicative of the redistribution of cells from the circulation to different tissues and organs as cell proliferation is unlikely, given that the effects of sleep already emerge within 3 h. In contrast, hematopoietic cell proliferation and differentiation take significantly longer to become evident [65, 79]. This rise and fall of the circulating leukocyte subsets across wake and sleep could reflect a homeostatic response and be maintained by the in and out of various tissues. Although this contention has not been systematically explored, some studies in rodents reported contrasting changes in blood versus other organs including lymph nodes, spleen, or bone marrow in sleep manipulation [75, 80–83]. The variability in the results from different sleep experiments in human and animal models could not allow for drawing a unifying picture of leukocyte distributions across sleep–wake states. These variabilities may partly be due to methodological issues; for example, differences in sleep manipulation and selection of only one or a few tissues in parallel. Moreover, the redistribution kinetics of leukocytes in humans and animals could be the function of their body size. Consequently, sleep appears to modulate the migration of leukocyte subsets across the tissues, the question of where these cells redistribute during sleep and why it is so remains largely open. There is the possibility that sleep modulates the inflammation homeostasis and that is reflected in the migration of leukocytes across the tissues.

4.2 Cytokines

Cytokines are extracellular proteins or glycoproteins that act as cell signaling molecules and are produced by nucleated cells [84]. These signaling molecules function in inflammation, immune response, and other physiological process, and communicate through autocrine, paracrine, and endocrine action at very low concentrations, such as at picomolar levels [53, 85]. A subset of cytokines is called chemokines. Chemokines actively participate in the inflammatory response attracting immune cells to the site of inflammation. Some reports suggest that cytokines can modulate sleep homeostasis and infection-mediated fatigue and sleepiness [86, 87]. Acute phase cytokines such as IL-6, tumor necrosis factor (TNF), and IL-1 are widely explored in the context of sleep. However, few studies report the role of other cytokines such as IL-2 and IL-12 (essential for adaptive immunity), and the anti-inflammatory cytokine IL-10, or the Th2 cytokine IL-4 in sleep modulation [53].

1. ***IL-1***. The IL-1 family of cytokines is classified into 11 classes having both analogous and different effects [88]. Most of the IL-1 is pro-inflammatory but IL-37 and IL-1RA have anti-inflammatory functions [89]. One night of sleep deprivation increases plasma IL-1β in healthy human subjects. In contrast, other studies subjected to either two nights of total sleep deprivation or four nights of REM sleep deprivation did not change it [69, 90, 91]. These mixed findings were also reported in rodent studies as some experiments reported a rise in IL-1 whereas others failed to substantiate these findings [63, 80, 92–95]. Like the plasma level of IL-1, mRNA expression also showed inconsistency in human and animal studies. Some acute and prolonged sleep manipulation in healthy human subjects reported no change in IL-1 production and mRNA expression [65, 71, 96]. At the same time, an increase in IL-1β mRNA levels was reported after 5 nights of sleep

restrictions in 13 healthy young men [97]. Animal studies measuring mRNA expression in peripheral blood lymphocytes and the spleen, liver, CNS, heart, and adipose tissue also reported increased levels of IL-1 [93, 98–102]. However, other animal studies reported no change [99, 103–106] or reduced levels of IL-1 in different tissues other than the blood following sleep manipulation [95, 103]. Moreover, rodent studies demonstrated a diurnal variation of IL-1β expression and protein abundance in the cerebral cortex having a peak during times of higher sleep propensity, i.e. early light period [107]. The higher IL-1β expression during the prolongation of wakefulness indicates that this diurnal variation might be largely attributed to increased local brain area use [108]. Furthermore, both acute and chronic loss of sleep enhances the expression of IL-1β not only in the cortex but also in several other brain areas, peripheral tissue, and circulation [85, 108, 109]. These findings suggest a potential role of IL-1 in regulating the sleep–wake cycle.

2. ***IL-6***. One night of either total or partial sleep deprivation has shown to increase, decrease, or leave plasma/salivary IL-6 levels unalteredry [65, 79, 90, 94, 110–117]. More consistent observations were reported following more than one night of sleep restrictions: plasma IL-6 or CRP (C-reactive protein) (hepatocytes synthesize CRP in response to IL-6) levels were mostly increased in such conditions [97, 118–120]. Similar results were also reported from the prolonged sleep deprivation experiments in the animals [93, 94, 121], although some studies did not find any change in plasma IL-6 or CRP levels both in humans and animals [66, 69, 92, 119, 122, 123], indicating this effect is transient and therefore needs to measure at multiple time points.

   One night of partial sleep deprivation increases intracellular IL-6 production in healthy humans [124–127]. A similar result was also found at the mRNA level, measured in peripheral blood lymphocytes of mice after 36 h of sleep deprivation [93]. Again, prolonged sleep restrictions even in humans increased intracellular IL-6 production [97, 128]. IL-6 protein or mRNA expression in various tissues including the brain is found to increase or have no change following sleep deprivations in animal [101, 104–106, 129, 130].
3. ***TNF***. The TNF family of cytokines is classified into 19 classes and 29 related receptors [131]. Among the TNF family, TNF-α is the most widely studied in the context of sleep and its function as a sleep regulator is well-established [132]. This cytokine is produced by nucleated cells including glia and neurons [133]. Like IL-1β, plasma TNF-α shows a mixed response to sleep. The acute sleep deprivation/restrictions did not change the plasma level of TNF in human subjects [69, 112, 113, 115, 119, 123]. Contrary to these findings, Dimitrov et al. [134] reported serum TNF levels reduced during sleep, and the following day circulating TNF showed a reverse pattern with higher levels after regular sleep than after the nocturnal vigil. Human studies of prolonged sleep restrictions reported no change in plasma TNF [69, 119, 123], however, a week of reduced sleep duration increased plasma TNF in one study [120]. Similarly, some rodent studies reported an increase in plasma TNF levels, whereas some other studies reported unchanged TNF plasma levels after sleep restriction [92, 93, 121, 135].

Animal studies have also measured the TNF protein or mRNA levels in different tissues under different sleep manipulations making it difficult to compare the results. Overall, most of the studies reported an increase, and some reported unchanged levels of TNF in the spleen, liver, brain, adipose tissue, and peritoneum after sleep loss [98, 99, 101, 103–106, 109, 129, 130, 136]. A few studies in humans also investigated the effect of experimental sleep manipulation on soluble TNF receptors I and II, which act as inhibitors and activators of TNF signaling. Acute sleep deprivation could not affect the plasma levels of these receptors, but prolonged sleep deprivation increases the levels of the soluble TNF receptor I which may be a response to the increase of TNF production by leukocytes [110, 119, 137].

4. ***Other cytokines***. Cytokines other than IL-1, TNF-α, and IL-6 are less investigated in the context of sleep. The experiments that measured the plasma IL-2 level failed to report the effect of sleep loss both in animal and human subjects [69, 92, 95, 122]. Some attempts were made to measure the anti-inflammatory cytokine IL-10 and the Th2 cytokine IL-4 in plasma. Both human and animal studies did not find any change in plasma IL-10 and IL-4 following sleep manipulation [69, 80, 95, 119, 121, 122, 138].

4.3 Proliferation and Activity of Immune Cells

Besides the studies of cytokines production, proliferation of lymphocytes and cytotoxicity of NK cells were also investigated in some studies during sleep manipulation. Acute sleep loss in human subjects decreases mitogen-stimulated lymphocyte proliferation whereas several nights with restricted sleep increase the peripheral blood mononuclear cells (PBMC) proliferation [74, 91, 97, 139]. On that line, reductions in lymphocyte proliferation were reported after 48 and 72 h of sleep deprivation in rats [140]. Other animal studies with chronic sleep restrictions paradigms failed to find such an effect [141, 142]. Similarly, the impact of sleep loss on NK-cell activity is complex. Acute sleep restriction reduces NK-cell activity in healthy human subjects [72, 91, 143]. Interestingly, some studies reported increased activity of NK cells in the evening after two nights without sleep; this could be a possible counterbalancing response to NK-cell activity reductions reported after acute sleep loss [72, 91, 143]. Reduction of cytotoxicity of NK cells after 24 and 48 h of sleep deprivation was reported in mice; interestingly, this effect was not evident after 72 h [144]. These data indicate that sleep loss influences the proliferation and activity of immune cells based on duration and type of interventions.

Some studies investigated the neutrophil function based on degranulation markers. A human study reported a transient increase in neutrophil degranulation after one, but not two nights without sleep [78]. Further, sleep improves the anti-bacterial action of neutrophils and monocytes in mice [145]. Recently, Sang et al. [146] have shown that prolonged sleep restriction increases the levels of circulating neutrophils with degranulation in mice.

4.4 Complement System

The complement system and its activation have yet to be much explored in the context of sleep. A human study has shown increased complement factors C3 and C5 after acute sleep deprivation [147]. Some studies did not find any change in C3, C4, or the receptors C3aR and C5aR after either acute total sleep deprivation or prolonged REM sleep deprivation [69, 148]. A recent study in mice suggests chronic sleep loss induces astrocyte-derived complement C3 secretion and microglial C3a receptor expression [149].

5 Immune Challenges and Sleep

Several immune-to-brain signaling pathways propagate the inflammatory response to the CNS following the activation of peripheral immune systems [150]. This peripheral–CNS crosstalk manifests as the behavioral and physiological changes considered symptoms of sickness, such as fatigue, lack of appetite, bad mood, and sleepiness [151]. These changes are the adaptive response of CNS to combat the infection or other aseptic insults. The inflammatory mediators that signal the CNS for these responses are cytokines and PGs which also modulate spontaneous sleep and sleepiness following infectious or inflammatory challenges. It is argued that PAMPs and DAMPs may act as upstream signals of these sleep-regulatory molecules. Identification of muramyl peptide as an upstream signal of sleep-regulatory substances and reduced NREM sleep in rats treated with antibiotics targeting gut bacteria strengthen the notion that gut microbiota and corresponding microbiota-derived PAMPs could be the possible regulators of sleep–wake behavior [152, 153]. Apart from microbiota-derived PAMPs, low levels of PAMPs are recognized by the daily interaction of our immune system with numerous microbes could trigger subtle increases in systemic cytokine and PG levels. The mechanistic details of PAMPs-mediated change in sleep are not clearly understood yet. Like PAMPs, DAMPs also initiate the activation of the immune system which could mediate the central nervous processes. DAMPs are released in extracellular space not only during injuries and cellular death but also during regular cellular activity. For example, synaptic firing, acute mental and physical stress generate DAMPs in the brain and periphery [154–156]. Furthermore, it is reported that sleep deprivation increases DAMPs in the brain, CSF, and blood [157–159]. Both these molecular processes activate the immune system through binding to respective PRRs, which likely leads to systemic increases in sleep-regulatory substances [160, 161].

Among sleep-regulatory cytokines IL-1 and TNF are widely explored. It is documented that IL-1 and TNF act as homeostatic NREM sleep regulators. Blocking the action of either IL-1 or TNF reduced NREM sleep duration and sleep rebound after sleep deprivation [162, 163]. Other cytokines have also been studied in this context.

Overall, a general understanding is that pro-inflammatory cytokines promote NREM sleep, whereas anti-inflammatory cytokines reduce NREM sleep [76].

LPS (lipopolysaccharides) and other infectious agents and their components increase IL-1β levels, and it also affects NREM sleep. Both central and peripheral applications of IL-1β increase NREM sleep in all species [53, 85]. The receptors of IL-1β are present in various cell types that include neurons, glia, epithelial cells, endothelial cells, macrophages, monocytes, neutrophils, and T lymphocytes [89, 164]. IL-1 binding to its receptors leads to myeloid differentiation primary response 88 (MYD88) signaling including c-Jun N-terminal kinase (JNK), p38 mitogen-activated protein kinase (MAPK), and NF-κB [89, 165] (Fig. 2). NF-κB is a

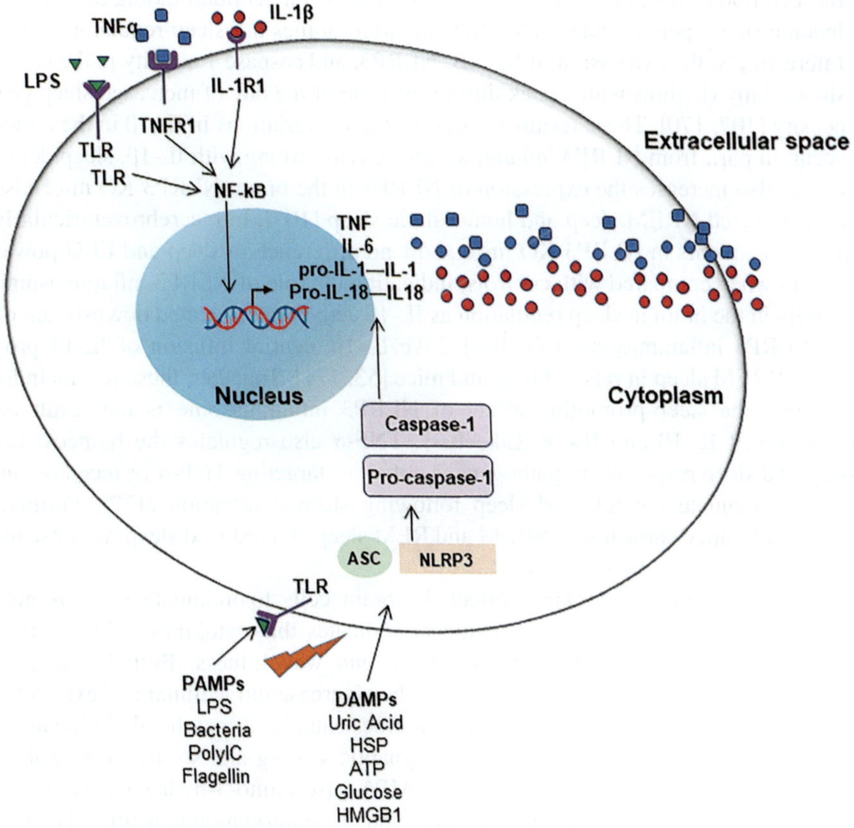

Fig. 2 Somnogenic cytokines signaling. The PAMPs and DAMPs action leads to the release of cytokines as these stimuli activate inflammatory signaling pathways such as NF-kB and inflammasome NLRP3. NF-kB stimulation results in the expression of pro-inflammatory cytokines TNF, IL-6, pro-IL-1, and pro-IL-18. The NLRP3 recruits pro-caspase 1 via ASC and converts it into caspase 1 which then pro-forms of IL-1b and IL-18 into their mature active forms. *ASC* apoptosis-associated speck-like protein containing a caspase recruit domain, *PAMPs* pathogen-associated molecular patterns, *DAMPs* danger/damage-associated molecular patterns, *NF-kB* nuclear factor-kappa B, *TNF* tumor necrosis factor, *IL* interleukin. Adapted from refs. [53, 174]

transcription factor that induces the expression of cytokines, chemokines, and other molecules that regulate sleep and immunity including IL-1β and TNF-α [53, 166]. At the same time, activation of TNF-α and IL-1β receptors signaling results in NF-κB transcription [166]. Further, NF-κB can also influence mitochondrial function [167]. NF-κB regulates a set of intracellular multi-protein complexes called inflammasomes. The components of inflammasome are assembled in response to PAMPs and DAMPs and characterized by the activation of caspase 1 [168]. Upon stimulation, the inflammasome receptors oligomerize and recruit pro-caspase 1 via apoptosis-associated speck-like protein containing a C-terminal caspase-recruitment domain (ASC) that leads conversion of pro-caspase 1 to active caspase 1. The activated caspase 1 converts pro-IL-1β and pro-IL-18 into their mature form leading to its secretion [169]. Among the inflammasomes, the nucleotide-binding domain and leucine-rich repeat protein-3 (NLRP3) inflammasomes are sleep regulatory [107]. Interestingly, the expression of NF-κB, NLRP3, and caspase 1 activity in the cortex shows daily rhythms with a peak during the time of the day of increased sleep propensity [107, 170]. These results suggest that daily variations in IL-1β in the cortex occur, in part, from NLRP3 inflammasome activity. Along with IL-1β, sleep deprivation also increases the expression of NLRP3 in the brain. NLRP3 KO mice also show reduced NREM sleep and homeostatic sleep [107]. Intracerebroventricularly IL-1β treatments in NLRP3 KO mice show no difference in sleep and EEG power spectra when compared with controls, indicating the role of NLRP3 inflammasome activity in the brain in sleep regulation as IL-1β activation is located downstream of the NLRP3 inflammasome [53, 107]. Like IL-1β, central infusion of IL-18 promotes NREM sleep in rats, rabbits, and mice [53, 171]. Together, these results indicate that the sleep-promoting effect of NLRP3 inflammasome is the result of induction of IL-1β and IL-18. Like IL-1, TNF-α also regulates the homeostatic sleep and sleep responses to pathogens. Antibodies targeting TNF-α or receptors in rodents attenuate the rebound sleep following sleep deprivation [172]. Further, TNFR1 KO mice show lower NREM and REM sleep and reduced sleep response to TNF-α [173].

How inflammatory cytokines affect the brain cells to modulate sleep is not explicitly known. However, some evidence indicates that cytokines influence the glutamate/GABA signaling crucial for sleep and wakefulness. Both IL-1β and TNF-α can induce glutamatergic activity as they increase intracellular and extracellular glutamate levels [175–178]. These cytokines can also affect the glial glutamatergic transmission. TNF-α can modulate synaptic scaling and so the sleep–wake state [179] and IL-1β modulates AMPA (α-amino-3-hydroxy-5-methyl-isoxazolepropionic acid) receptor expression and phosphorylation in neurons [180].

The cyclooxygenase (COX)-PG pathway is another inflammatory mechanism influencing sleep physiology. PGs are lipid mediators that are synthesized by the action of COX from arachidonic acid [181]. PGs mediate the signs of inflammation like fever and pain. The nonsteroidal anti-inflammatory drugs (NSAIDs), e.g., ibuprofen or acetylsalicylic acid (aspirin), primarily prevent the synthesis of PGs through the inhibition of COX enzymes [182]. Sleep-promoting effects of PGs were observed in animal studies where intracerebroventricular infusion of PGD2 substantially increased the NREM sleep [183, 184]. Further, inhibition of PG production by COX-2 inhibitors reduced spontaneous and TNFα-induced NREM sleep [185, 186]. Loss of sleep increases the CSF levels of PGD2, E2, and F2a in rats [187]. This indicates a loss of sleep may target COX enzymes, the precursor of all functional PGs, rather than a specific PG. More recently, Sang et al have shown that prolonged sleep disruption that witnesses cytokine-storm-like syndrome also increases levels of PGD2 in the brain and this leads to elevated PGD2 efflux across the blood-brain-barrier—mediated by ATP-binding cassette subfamily C4 transporter [146]. Disruption of PGD2-DP1 (prostaglandin D2 receptor 1) signaling reduces sleep-deprivation-induced inflammation, suggesting a sleep-related change in brain PGD2 levels leads to a profound impact on the peripheral immune system [146].

6 Neuro-Immune Crosstalk and Sleep

The CNS and immune system are independent and interacting entities. Both of these systems closely interact to sense and respond to the environmental stimuli and store these actions as memories in a way the organism would be prepared for similar stimulus re-encounters in the future for better adaptability. An example of mutual interaction is the induction of an inflammatory response by the activation of the hypothalamus–pituitary–adrenal (HPA) axis and the autonomic nervous system (ANS) [155, 188]. A plausible explanation for this phenomenon could be the idea that any kind of stressor (e.g., facing a predator or other environmental challenges) also activates the immune system along with the freeze, fight, or flight response to cope with the pathogen invading the wound and accelerate the healing process [174]. Reversely, immune system activation prompts neurobehavioral, neuroendocrine, and ANS responses [189, 190]. This process reflects the notion that the immune system functions as a "sixth sense", beyond taste, touch, smell, and hearing, conveying signals of microbial or other immunogenic encounters to the brain [191, 192]. As a defense measure, the CNS generates a set of measurable behavioral and physiological changes; for example, sleepiness, fever, and other sickness behaviors [151, 193].

The nerve fibers, leukocytes, and soluble mediators bridge the CNS and the immune system. Both primary and secondary lymphoid organs are innervated by the sympathetic, peptidergic, and sensory nerve fibers [194]. The efferent nerve neurotransmitters and neuropeptides like catecholamines or neuropeptide Y can be recognized by their respective receptors on immune cells, and leucocytes can synthesize and release these neuronal messengers [195–197]. The cholinergic innervation of immune organs is not well-established yet; however, the T cell cholinergic system can provide acetylcholine release in the periphery [198, 199]. Furthermore, tissue-resident leukocytes can also sense the neurotransmitters and neuropeptides released by local nerve endings [196]. In contrast to fast communication by nerve fiber, neuroendocrine signaling provides slower control over the immune system. Leukocytes can sense and synthesize neuroendocrine mediators [196, 200–203].

The ability of leukocytes to synthesize the neuromodulators allows them to send signals to the brain via nerve fiber or blood circulation. The cytokines and chemokines are mediators of immune–brain signaling, which are produced by the immune cells and have their respective receptors on neurons, astrocytes, and microglia. These molecules along with other inflammatory agents like PGs, PAMPs, and DAMPs can reach the peripheral nervous system (PNS) or CNS by crossing the blood-brain barrier via transporter systems or at leaky sites of circumventricular organs including the subfornical organ, area postrema, vascular organ of the lamina terminalis, median eminence, pituitary gland, and pineal gland [204, 205]. Other immune–brain signaling processes involve vagal or other neural afferents, as well as cerebral endothelial and perivascular cells or the directed traffic of leukocytes to the brain [206–209]. These pathways could contribute to microglial activation and neuroinflammation in response to peripheral inflammation [210, 211]. The vagal nerve is an important part of PNS having control of the viscera and mediating oxygen demand by altering respiratory control of the diaphragm, lungs, and heart [212]. Stimulation of vagal afferent transduces the signal from visceral organs to the dorsal vagal complex which involves the nucleus tractus solitarius (NTS), the dorsal motor nucleus (DMN), and the area postrema that is in the medulla area of the brainstem [53] (Fig. 3). The NTS projects to several brain areas including sleep-regulating areas like the cortex, hypothalamus, nucleus accumbens, and brainstem [213]. Vagal efferents transmit signals from the brainstem to peripheral organs and its stimulation reduces inflammation [214]. In rodents, vagotomy attenuates the NREM sleep and SWA following IL-1β, TNF-α, or LPS treatment [215–217]. Peripheral treatment of pro-inflammatory molecules IL-1b and TNF-α or LPS increases the expression of IL-1β and TNF-α expression in the cortex, hypothalamus, and NTS by stimulating the vagal afferents [217–219]. At the same time, vagotomies attenuate IL-1β and TNF-α expression in the brain after IL-1β, TNF-α, or LPS is applied to the peritoneum. These results suggest that peripheral inflammation stimulates brain inflammatory molecules through the vagal afferents. Interestingly, higher levels of peripheral inflammatory stimuli reduce the effects of the vagotomy, suggesting that inflammatory molecules are increased in the brain through leaky areas of the blood-brain-barrier in the circumventricular organs [214, 217, 220]. Thus, sleep modularly effects of the vagal afferents occur, in part, through the translation of inflammatory signals between the brain and periphery. Further, the synthesis of chemokines and cytokines by the brain cells and nerve fibers could demonstrate the pathway of brain-to-immune signaling [206, 221]. These results show overlapping biochemical mechanisms affecting intra- and intersystem communication between the CNS and the immune system.

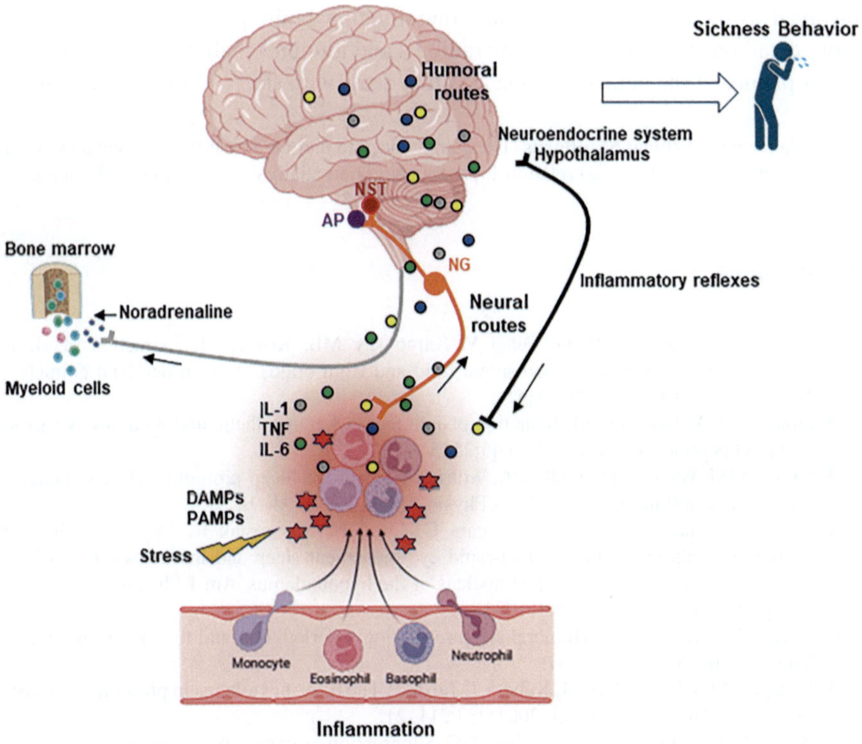

Fig. 3 Schematic illustration of immune–brain crosstalk. The stressor-activated sympathetic nervous system fibers resulting in the release of noradrenaline which induces the production of myeloid cells. The action of DAMPs and PAMPs on these cells releases pro-inflammatory cytokines that access the brain via neural humoral routes. This leads to the brain's integration of inflammatory signaling and controlling autonomic, neuroendocrine, and behavioral responses (e.g., sickness behavior). *PAMPs* pathogen-associated molecular patterns, *DAMPs* danger/damage-associated molecular patterns, *TNF* tumor necrosis factor, *IL* interleukin, *AP* area postrema, *NTS* nucleus tractus solitarius, *NG* nodose ganglion. Adapted from refs. [174, 222]

7 Conclusion

Sleep and the immune system are bidirectionally linked. The notion that sleep is important for immune functions existed for a long time. Loss of sleep disrupts a wide array of immune parameters. At the same, several molecules and pathways associated with the immune system can modulate spontaneous, physiological sleep, and sickness-mediated sleep when the animal's host defense is compromised by pathogen invasions. This interaction between sleep and the immune system is a relatively new research avenue. Therefore, details of how the immune-to-sleep and sleep-to-immune axis work are not well understood. Various parameters are likely involved to mediate these functions. Investigating brain–periphery cross-talks in more detail during different sleep paradigms and immune challenges would bring a better picture of the neuro-immune axis that deals with sleep–immune interactions.

Compliance with Ethical Standards The authors declare that the research was conducted without any commercial or financial relationships that could potentially create a conflict of interest.

References

1. Pappenheimer JR, Koski G, Fencl V, Karnovsky ML, Krueger J. Extraction of sleep-promoting factor S from cerebrospinal fluid and from brains of sleep-deprived animals. J Neurophysiol. 1975;38:1299–311.
2. Dinarello CA, Krueger JM. Induction of interleukin 1 by synthetic and naturally occurring muramyl peptides. Fed Proc. 1986;45:2545–8.
3. Krueger JM, Walter J, Dinarello CA, Wolff SM, Chedid L. Sleep-promoting effects of endogenous pyrogen (interleukin-1). Am J Physiol. 1984;246:R994–9.
4. Baker FC, Shah S, Stewart D, Angara C, Gong H, Szymusiak R, Opp MR, McGinty D. Interleukin 1beta enhances non-rapid eye movement sleep and increases c-Fos protein expression in the median preoptic nucleus of the hypothalamus. Am J Physiol Regul Integr Comp Physiol. 2005;288:R998–1005.
5. Jewett KA, Krueger JM. Humoral sleep regulation; interleukin-1 and tumor necrosis factor. Vitam Horm. 2012;89:241–57.
6. Krueger JM, Obal FJ, Fang J, Kubota T, Taishi P. The role of cytokines in physiological sleep regulation. Ann N Y Acad Sci. 2001;933:211–21.
7. Brown R, Pang G, Husband AJ, King MG. Suppression of immunity to influenza virus infection in the respiratory tract following sleep disturbance. Reg Immunol. 1989;2:321–5.
8. Imeri L, Opp MR. How (and why) the immune system makes us sleep. Nat Rev Neurosci. 2009;10:199–210.
9. Opp MR, Krueger JM. Interleukin 1-receptor antagonist blocks interleukin 1-induced sleep and fever. Am J Physiol. 1991;260:R453–7.
10. Aschoff J. Circadian rhythms in man. Science. 1965;148:1427–32.
11. Pittendrigh CS. Circadian rhythms and the circadian organization of living systems. Cold Spring Harb Symp Quant Biol. 1960;25:159–84.
12. Dibner C, Schibler U, Albrecht U. The mammalian circadian timing system: organization and coordination of central and peripheral clocks. Annu Rev Physiol. 2010;72:517–49.
13. Golombek DA, Rosenstein RE. Physiology of circadian entrainment. Physiol Rev. 2010;90:1063–102.
14. Kalsbeek A, Palm IF, La Fleur SE, Scheer FA, Perreau-Lenz S, Ruiter M, Kreier F, Cailotto C, Buijs RM. SCN outputs and the hypothalamic balance of life. J Biol Rhythms. 2006;21:458–69.
15. Takahashi JS. Finding new clock components: past and future. J Biol Rhythms. 2004;19:339–47.
16. Welsh DK, Takahashi JS, Kay SA. Suprachiasmatic nucleus: cell autonomy and network properties. Annu Rev Physiol. 2010;72:551–77.
17. Lucas RJ, Lall GS, Allen AE, Brown TM. How rod, cone, and melanopsin photoreceptors come together to enlighten the mammalian circadian clock. Prog Brain Res. 2012;199:1–18.
18. Moore RY, Lenn NJ. A retinohypothalamic projection in the rat. J Comp Neurol. 1972;146:1–14.
19. Mohawk JA, Green CB, Takahashi JS. Central and peripheral circadian clocks in mammals. Annu Rev Neurosci. 2012;35:445–62.

20. Moore RY. Suprachiasmatic nucleus in sleep-wake regulation. Sleep Med. 2007;8(Suppl 3):27–33.
21. Weaver DR. The suprachiasmatic nucleus: a 25-year retrospective. J Biol Rhythms. 1998;13:100–12.
22. Vadnie CA, McClung CA. Circadian rhythm disturbances in mood disorders: insights into the role of the suprachiasmatic nucleus. Neural Plast. 2017;2017:1504507.
23. Fekete M, van Ree JM, Niesink RJ, de Wied D. Disrupting circadian rhythms in rats induces retrograde amnesia. Physiol Behav. 1985;34:883–7.
24. Folkard S, Wever RA, Wildgruber CM. Multi-oscillatory control of circadian rhythms in human performance. Nature. 1983;305:223–6.
25. Ruby NF, Hwang CE, Wessells C, Fernandez F, Zhang P, Sapolsky R, Heller HC. Hippocampal-dependent learning requires a functional circadian system. Proc Natl Acad Sci U S A. 2008;105:15593–8.
26. Buhr ED, Takahashi JS. Molecular components of the Mammalian circadian clock. Handb Exp Pharmacol. 2013;3–27
27. Ko CH, Takahashi JS. Molecular components of the mammalian circadian clock. Hum Mol Genet. 2006;15 Spec No 2:R271–7.
28. Liu AC, Welsh DK, Ko CH, Tran HG, Zhang EE, Priest AA, Buhr ED, Singer O, Meeker K, Verma IM, Doyle FJ 3rd, Takahashi JS, Kay SA. Intercellular coupling confers robustness against mutations in the SCN circadian clock network. Cell. 2007;129:605–16.
29. Crumbley C, Burris TP. Direct regulation of CLOCK expression by REV-ERB. PLoS One. 2011;6:e17290.
30. Reppert SM, Weaver DR. Coordination of circadian timing in mammals. Nature. 2002;418:935–41.
31. King DP, Zhao Y, Sangoram AM, Wilsbacher LD, Tanaka M, Antoch MP, Steeves TD, Vitaterna MH, Kornhauser JM, Lowrey PL, Turek FW, Takahashi JS. Positional cloning of the mouse circadian clock gene. Cell. 1997;89:641–53.
32. Zheng B, Albrecht U, Kaasik K, Sage M, Lu W, Vaishnav S, Li Q, Sun ZS, Eichele G, Bradley A, Lee CC. Nonredundant roles of the mPer1 and mPer2 genes in the mammalian circadian clock. Cell. 2001;105:683–94.
33. Duong HA, Robles MS, Knutti D, Weitz CJ. A molecular mechanism for circadian clock negative feedback. Science. 2011;332:1436–9.
34. Gallego M, Virshup DM. Post-translational modifications regulate the ticking of the circadian clock. Nat Rev Mol Cell Biol. 2007;8:139–48.
35. Akashi M, Takumi T. The orphan nuclear receptor RORalpha regulates circadian transcription of the mammalian core-clock Bmal1. Nat Struct Mol Biol. 2005;12:441–8.
36. Preitner N, Damiola F, Lopez-Molina L, Zakany J, Duboule D, Albrecht U, Schibler U. The orphan nuclear receptor REV-ERBalpha controls circadian transcription within the positive limb of the mammalian circadian oscillator. Cell. 2002;110:251–60.
37. Guillaumond F, Dardente H, Giguere V, Cermakian N. Differential control of Bmal1 circadian transcription by REV-ERB and ROR nuclear receptors. J Biol Rhythms. 2005;20:391–403.
38. Sato TK, Panda S, Miraglia LJ, Reyes TM, Rudic RD, McNamara P, Naik KA, FitzGerald GA, Kay SA, Hogenesch JB. A functional genomics strategy reveals Rora as a component of the mammalian circadian clock. Neuron. 2004;43:527–37.
39. Feinberg I, March JD, Fein G, Floyd TC, Walker JM, Price L. Period and amplitude analysis of 0.5-3 c/sec activity in NREM sleep of young adults. Electroencephalogr Clin Neurophysiol. 1978;44:202–13.
40. Borbely AA. A two process model of sleep regulation. Hum Neurobiol. 1982;1:195–204.
41. Deboer T. Behavioral and electrophysiological correlates of sleep and sleep homeostasis. Curr Top Behav Neurosci. 2015;25:1–24.
42. Radulovacki M, Virus RM, Djuricic-Nedelson M, Green RD. Adenosine analogs and sleep in rats. J Pharmacol Exp Ther. 1984;228:268–74.

43. Scharf MT, Naidoo N, Zimmerman JE, Pack AI. The energy hypothesis of sleep revisited. Prog Neurobiol. 2008;86:264–80.
44. Gallo RL, Hooper LV. Epithelial antimicrobial defence of the skin and intestine. Nat Rev Immunol. 2012;12:503–16.
45. Bianchi ME. DAMPs, PAMPs and alarmins: all we need to know about danger. J Leukoc Biol. 2007;81:1–5.
46. Iwasaki A, Medzhitov R. Control of adaptive immunity by the innate immune system. Nat Immunol. 2015;16:343–53.
47. Matzinger P. Friendly and dangerous signals: is the tissue in control? Nat Immunol. 2007;8:11–3.
48. Vidarsson G, Dekkers G, Rispens T. IgG subclasses and allotypes: from structure to effector functions. Front Immunol. 2014;5:520.
49. Chovatiya R, Medzhitov R. Stress, inflammation, and defense of homeostasis. Mol Cell. 2014;54:281–8.
50. Kotas ME, Medzhitov R. Homeostasis, inflammation, and disease susceptibility. Cell. 2015;160:816–27.
51. Medzhitov R. Origin and physiological roles of inflammation. Nature. 2008;454:428–35.
52. Sevenich L. Brain-resident microglia and blood-borne macrophages orchestrate central nervous system inflammation in neurodegenerative disorders and brain cancer. Front Immunol. 2018;9:697.
53. Zielinski MR, Gibbons AJ. Neuroinflammation, sleep, and circadian rhythms. Front Cell Infect Microbiol. 2022;12:853096.
54. Amini M, Yousefi Z, Ghafori SS, Hassanzadeh G. Sleep deprivation and NLRP3 inflammasome: Is there a causal relationship? Front Neurosci. 2022;16:1018628.
55. Garbarino S, Lanteri P, Bragazzi NL, Magnavita N, Scoditti E. Role of sleep deprivation in immune-related disease risk and outcomes. Commun Biol. 2021;4:1304.
56. Periasamy S, Hsu DZ, Fu YH, Liu MY. Sleep deprivation-induced multi-organ injury: role of oxidative stress and inflammation. EXCLI J. 2015;14:672–83.
57. Cevenini E, Monti D, Franceschi C. Inflamm-ageing. Curr Opin Clin Nutr Metab Care. 2013;16:14–20.
58. Franceschi C, Bonafe M, Valensin S, Olivieri F, De Luca M, Ottaviani E, De Benedictis G. Inflamm-aging. An evolutionary perspective on immunosenescence. Ann N Y Acad Sci. 2000;908:244–54.
59. Boteanu RM, Suica VI, Uyy E, Ivan L, Dima SO, Popescu I, Simionescu M, Antohe F. Alarmins in chronic noncommunicable diseases: Atherosclerosis, diabetes and cancer. J Proteomics. 2017;153:21–9.
60. Chen GY, Nunez G. Sterile inflammation: sensing and reacting to damage. Nat Rev Immunol. 2010;10:826–37.
61. Camell C, Goldberg E, Dixit VD. Regulation of Nlrp3 inflammasome by dietary metabolites. Semin Immunol. 2015;27:334–42.
62. Geiger SS, Fagundes CT, Siegel RM. Chrono-immunology: progress and challenges in understanding links between the circadian and immune systems. Immunology. 2015;146:349–58.
63. Scheiermann C, Kunisaki Y, Frenette PS. Circadian control of the immune system. Nat Rev Immunol. 2013;13:190–8.
64. Casanova-Acebes M, Nicolas-Avila JA, Li JL, Garcia-Silva S, Balachander A, Rubio-Ponce A, Weiss LA, Adrover JM, Burrows K, A-González N, Ballesteros I, Devi S, Quintana JA, Crainiciuc G, Leiva M, Gunzer M, Weber C, Nagasawa T, Soehnlein O, Merad M, Mortha A, Ng LG, Peinado H, Hidalgo A. Neutrophils instruct homeostatic and pathological states in naive tissues. J Exp Med. 2018;215:2778–95.
65. Born J, Lange T, Hansen K, Molle M, Fehm HL. Effects of sleep and circadian rhythm on human circulating immune cells. J Immunol. 1997;158:4454–64.

66. Boudjeltia KZ, Faraut B, Stenuit P, Esposito MJ, Dyzma M, Brohee D, Ducobu J, Vanhaeverbeek M, Kerkhofs M. Sleep restriction increases white blood cells, mainly neutrophil count, in young healthy men: a pilot study. Vasc Health Risk Manag. 2008;4:1467–70.
67. Lasselin J, Rehman JU, Akerstedt T, Lekander M, Axelsson J. Effect of long-term sleep restriction and subsequent recovery sleep on the diurnal rhythms of white blood cell subpopulations. Brain Behav Immun. 2015;47:93–9.
68. Liu H, Wang G, Luan G, Liu Q. Effects of sleep and sleep deprivation on blood cell count and hemostasis parameters in healthy humans. J Thromb Thrombolysis. 2009;28:46–9.
69. Ruiz FS, Andersen ML, Martins RC, Zager A, Lopes JD, Tufik S. Immune alterations after selective rapid eye movement or total sleep deprivation in healthy male volunteers. Innate Immun. 2012;18:44–54.
70. Ackermann K, Revell VL, Lao O, Rombouts EJ, Skene DJ, Kayser M. Diurnal rhythms in blood cell populations and the effect of acute sleep deprivation in healthy young men. Sleep. 2012;35:933–40.
71. Chennaoui M, Arnal PJ, Drogou C, Leger D, Sauvet F, Gomez-Merino D. Leukocyte expression of type 1 and type 2 purinergic receptors and pro-inflammatory cytokines during total sleep deprivation and/or sleep extension in healthy subjects. Front Neurosci. 2017;11:240.
72. Irwin M, McClintick J, Costlow C, Fortner M, White J, Gillin JC. Partial night sleep deprivation reduces natural killer and cellular immune responses in humans. FASEB J. 1996;10:643–53.
73. Ricardo JS, Cartner L, Oliver SJ, Laing SJ, Walters R, Bilzon JL, Walsh NP. No effect of a 30-h period of sleep deprivation on leukocyte trafficking, neutrophil degranulation and saliva IgA responses to exercise. Eur J Appl Physiol. 2009;105:499–504.
74. Wilder-Smith A, Mustafa FB, Earnest A, Gen L, Macary PA. Impact of partial sleep deprivation on immune markers. Sleep Med. 2013;14:1031–4.
75. Guariniello LD, Vicari P, Lee KS, de Oliveira AC, Tufik S. Bone marrow and peripheral white blood cells number is affected by sleep deprivation in a murine experimental model. J Cell Physiol. 2012;227:361–6.
76. Besedovsky L, Lange T, Haack M. The sleep-immune crosstalk in health and disease. Physiol Rev. 2019;99:1325–80.
77. Kerkhofs M, Boudjeltia KZ, Stenuit P, Brohee D, Cauchie P, Vanhaeverbeek M. Sleep restriction increases blood neutrophils, total cholesterol and low density lipoprotein cholesterol in postmenopausal women: A preliminary study. Maturitas. 2007;56:212–5.
78. Costa RJ, Smith AH, Oliver SJ, Walters R, Maassen N, Bilzon JL, Walsh NP. The effects of two nights of sleep deprivation with or without energy restriction on immune indices at rest and in response to cold exposure. Eur J Appl Physiol. 2010;109:417–28.
79. Dimitrov S, Lange T, Benedict C, Nowell MA, Jones SA, Scheller J, Rose-John S, Born J. Sleep enhances IL-6 trans-signaling in humans. FASEB J. 2006;20:2174–6.
80. Everson CA. Clinical assessment of blood leukocytes, serum cytokines, and serum immunoglobulins as responses to sleep deprivation in laboratory rats. Am J Physiol Regul Integr Comp Physiol. 2005;289:R1054–63.
81. Lungato L, Nogueira-Pedro A, Carvalho Dias C, Paredes-Gamero EJ, Tufik S, D'Almeida V. Effects of sleep deprivation on mice bone marrow and spleen B lymphopoiesis. J Cell Physiol. 2016;231:1313–20.
82. Zager A, Andersen ML, Ruiz FS, Antunes IB, Tufik S. Effects of acute and chronic sleep loss on immune modulation of rats. Am J Physiol Regul Integr Comp Physiol. 2007;293:R504–9.
83. Zager A, Ruiz FS, Tufik S, Andersen ML. Immune outcomes of paradoxical sleep deprivation on cellular distribution in naive and lipopolysaccharide-stimulated mice. Neuroimmunomodulation. 2012;19:79–87.
84. Oppenheim JJ. Cytokines: past, present, and future. Int J Hematol. 2001;74:3–8.
85. Zielinski MR, Krueger JM. Sleep and innate immunity. Front Biosci (Schol Ed). 2011;3:632–42.
86. Krueger JM. The role of cytokines in sleep regulation. Curr Pharm Des. 2008;14:3408–16.
87. Krueger JM, Rector DM, Churchill L. Sleep and cytokines. Sleep Med Clin. 2007;2:161–9.

88. Xu D, Mu R, Wei X. The roles of IL-1 family cytokines in the pathogenesis of systemic sclerosis. Front Immunol. 2019;10:2025.
89. Dinarello CA. Overview of the IL-1 family in innate inflammation and acquired immunity. Immunol Rev. 2018;281:8–27.
90. Frey DJ, Fleshner M, Wright KP Jr. The effects of 40 hours of total sleep deprivation on inflammatory markers in healthy young adults. Brain Behav Immun. 2007;21:1050–7.
91. Moldofsky H, Lue FA, Davidson JR, Gorczynski R. Effects of sleep deprivation on human immune functions. FASEB J. 1989;3:1972–7.
92. Hirotsu C, Rydlewski M, Araujo MS, Tufik S, Andersen ML. Sleep loss and cytokines levels in an experimental model of psoriasis. PLoS One. 2012;7:e51183.
93. Hu J, Chen Z, Gorczynski CP, Gorczynski LY, Kai Y, Lee L, Manuel J, Gorczynski RM. Sleep-deprived mice show altered cytokine production manifest by perturbations in serum IL-1ra, TNFa, and IL-6 levels. Brain Behav Immun. 2003;17:498–504.
94. Thimgan MS, Gottschalk L, Toedebusch C, McLeland J, Rechtschaffen A, Gilliland-Roberts M, Duntley SP, Shaw PJ. Cross-translational studies in human and Drosophila identify markers of sleep loss. PLoS One. 2013;8:e61016.
95. Trammell RA, Toth LA. Effects of sleep fragmentation and chronic latent viral infection on behavior and inflammation in mice. Comp Med. 2015;65:173–85.
96. Axelsson J, Rehman JU, Akerstedt T, Ekman R, Miller GE, Hoglund CO, Lekander M. Effects of sustained sleep restriction on mitogen-stimulated cytokines, chemokines and T helper 1/ T helper 2 balance in humans. PLoS One. 2013;8:e82291.
97. van Leeuwen WM, Lehto M, Karisola P, Lindholm H, Luukkonen R, Sallinen M, Harma M, Porkka-Heiskanen T, Alenius H. Sleep restriction increases the risk of developing cardiovascular diseases by augmenting proinflammatory responses through IL-17 and CRP. PLoS One. 2009;4:e4589.
98. Ashley NT, Sams DW, Brown AC, Dumaine JE. Novel environment influences the effect of paradoxical sleep deprivation upon brain and peripheral cytokine gene expression. Neurosci Lett. 2016;615:55–9.
99. Dumaine JE, Ashley NT. Acute sleep fragmentation induces tissue-specific changes in cytokine gene expression and increases serum corticosterone concentration. Am J Physiol Regul Integr Comp Physiol. 2015;308:R1062–9.
100. Kang WS, Park HJ, Chung JH, Kim JW. REM sleep deprivation increases the expression of interleukin genes in mice hypothalamus. Neurosci Lett. 2013;556:73–8.
101. Kincheski GC, Valentim IS, Clarke JR, Cozachenco D, Castelo-Branco MTL, Ramos-Lobo AM, Rumjanek V, Donato J Jr, De Felice FG, Ferreira ST. Chronic sleep restriction promotes brain inflammation and synapse loss, and potentiates memory impairment induced by amyloid-beta oligomers in mice. Brain Behav Immun. 2017;64:140–51.
102. Mackiewicz M, Sollars PJ, Ogilvie MD, Pack AI. Modulation of IL-1 beta gene expression in the rat CNS during sleep deprivation. Neuroreport. 1996;7:529–33.
103. Ashley NT, Walton JC, Haim A, Zhang N, Prince LA, Fruchey AM, Lieberman RA, Weil ZM, Magalang UJ, Nelson RJ. Sleep deprivation attenuates endotoxin-induced cytokine gene expression independent of day length and circulating cortisol in male Siberian hamsters (Phodopus sungorus). J Exp Biol. 2013;216:2581–6.
104. Manchanda S, Singh H, Kaur T, Kaur G. Low-grade neuroinflammation due to chronic sleep deprivation results in anxiety and learning and memory impairments. Mol Cell Biochem. 2018;449:63–72.
105. Vacas S, Degos V, Maze M. Fragmented sleep enhances postoperative neuroinflammation but not cognitive dysfunction. Anesth Analg. 2017;124:270–6.
106. Zielinski MR, Davis JM, Fadel JR, Youngstedt SD. Influence of chronic moderate sleep restriction and exercise on inflammation and carcinogenesis in mice. Brain Behav Immun. 2012;26:672–9.
107. Zielinski MR, Gerashchenko D, Karpova SA, Konanki V, McCarley RW, Sutterwala FS, Strecker RE, Basheer R. The NLRP3 inflammasome modulates sleep and NREM sleep delta

power induced by spontaneous wakefulness, sleep deprivation and lipopolysaccharide. Brain Behav Immun. 2017;62:137–50.
108. Zielinski MR, McKenna JT, McCarley RW. Functions and mechanisms of sleep. AIMS Neurosci. 2016;3:67–104.
109. Zielinski MR, Kim Y, Karpova SA, McCarley RW, Strecker RE, Gerashchenko D. Chronic sleep restriction elevates brain interleukin-1 beta and tumor necrosis factor-alpha and attenuates brain-derived neurotrophic factor expression. Neurosci Lett. 2014;580:27–31.
110. Chennaoui M, Sauvet F, Drogou C, Van Beers P, Langrume C, Guillard M, Gourby B, Bourrilhon C, Florence G, Gomez-Merino D. Effect of one night of sleep loss on changes in tumor necrosis factor alpha (TNF-alpha) levels in healthy men. Cytokine. 2011;56:318–24.
111. Faraut B, Nakib S, Drogou C, Elbaz M, Sauvet F, De Bandt JP, Leger D. Napping reverses the salivary interleukin-6 and urinary norepinephrine changes induced by sleep restriction. J Clin Endocrinol Metab. 2015;100:E416–26.
112. Haack M, Schuld A, Kraus T, Pollmacher T. Effects of sleep on endotoxin-induced host responses in healthy men. Psychosom Med. 2001;63:568–78.
113. Irwin MR, Rinetti G. Disordered sleep, nocturnal cytokines, and immunity: interactions between alcohol dependence and African-American ethnicity. Alcohol. 2004;32:53–61.
114. Redwine L, Hauger RL, Gillin JC, Irwin M. Effects of sleep and sleep deprivation on interleukin-6, growth hormone, cortisol, and melatonin levels in humans. J Clin Endocrinol Metab. 2000;85:3597–603.
115. Sauvet F, Leftheriotis G, Gomez-Merino D, Langrume C, Drogou C, Van Beers P, Bourrilhon C, Florence G, Chennaoui M. Effect of acute sleep deprivation on vascular function in healthy subjects. J Appl Physiol (1985). 2010;108:68–75.
116. Vgontzas AN, Papanicolaou DA, Bixler EO, Lotsikas A, Zachman K, Kales A, Prolo P, Wong ML, Licinio J, Gold PW, Hermida RC, Mastorakos G, Chrousos GP. Circadian interleukin-6 secretion and quantity and depth of sleep. J Clin Endocrinol Metab. 1999;84:2603–7.
117. Vgontzas AN, Pejovic S, Zoumakis E, Lin HM, Bixler EO, Basta M, Fang J, Sarrigiannidis A, Chrousos GP. Daytime napping after a night of sleep loss decreases sleepiness, improves performance, and causes beneficial changes in cortisol and interleukin-6 secretion. Am J Physiol Endocrinol Metab. 2007;292:E253–61.
118. Haack M, Sanchez E, Mullington JM. Elevated inflammatory markers in response to prolonged sleep restriction are associated with increased pain experience in healthy volunteers. Sleep. 2007;30:1145–52.
119. Shearer WT, Reuben JM, Mullington JM, Price NJ, Lee BN, Smith EO, Szuba MP, Van Dongen HP, Dinges DF. Soluble TNF-alpha receptor 1 and IL-6 plasma levels in humans subjected to the sleep deprivation model of spaceflight. J Allergy Clin Immunol. 2001;107:165–70.
120. Vgontzas AN, Zoumakis E, Bixler EO, Lin HM, Follett H, Kales A, Chrousos GP. Adverse effects of modest sleep restriction on sleepiness, performance, and inflammatory cytokines. J Clin Endocrinol Metab. 2004;89:2119–26.
121. Yehuda S, Sredni B, Carasso RL, Kenigsbuch-Sredni D. REM sleep deprivation in rats results in inflammation and interleukin-17 elevation. J Interferon Cytokine Res. 2009;29:393–8.
122. Hurtado-Alvarado G, Becerril-Villanueva E, Contis-Montes de Oca A, Dominguez-Salazar E, Salinas-Jazmin N, Perez-Tapia SM, Pavon L, Velazquez-Moctezuma J, Gomez-Gonzalez B. The yin/yang of inflammatory status: Blood-brain barrier regulation during sleep. Brain Behav Immun. 2018;69:154–66.
123. Lekander M, Andreasson AN, Kecklund G, Ekman R, Ingre M, Akerstedt T, Axelsson J. Subjective health perception in healthy young men changes in response to experimentally restricted sleep and subsequent recovery sleep. Brain Behav Immun. 2013;34:43–6.
124. Carroll JE, Carrillo C, Olmstead R, Witarama T, Breen EC, Yokomizo M, Seeman T, Irwin MR. Sleep deprivation and divergent toll-like receptor-4 activation of cellular inflammation in aging. Sleep. 2015;38:205–11.
125. Irwin MR, Carrillo C, Olmstead R. Sleep loss activates cellular markers of inflammation: sex differences. Brain Behav Immun. 2010;24:54–7.

126. Irwin MR, Wang M, Campomayor CO, Collado-Hidalgo A, Cole S. Sleep deprivation and activation of morning levels of cellular and genomic markers of inflammation. Arch Intern Med. 2006;166:1756–62.
127. Irwin MR, Witarama T, Caudill M, Olmstead R, Breen EC. Sleep loss activates cellular inflammation and signal transducer and activator of transcription (STAT) family proteins in humans. Brain Behav Immun. 2015;47:86–92.
128. Simpson NS, Diolombi M, Scott-Sutherland J, Yang H, Bhatt V, Gautam S, Mullington J, Haack M. Repeating patterns of sleep restriction and recovery: Do we get used to it? Brain Behav Immun. 2016;58:142–51.
129. Kaur T, Singh H, Mishra R, Manchanda S, Gupta M, Saini V, Sharma A, Kaur G. Withania somnifera as a potential anxiolytic and immunomodulatory agent in acute sleep deprived female Wistar rats. Mol Cell Biochem. 2017;427:91–101.
130. Venancio DP, Suchecki D. Prolonged REM sleep restriction induces metabolic syndrome-related changes: Mediation by pro-inflammatory cytokines. Brain Behav Immun. 2015;47:109–17.
131. Dostert C, Grusdat M, Letellier E, Brenner D. The TNF family of ligands and receptors: communication modules in the immune system and beyond. Physiol Rev. 2019;99:115–60.
132. Rockstrom MD, Chen L, Taishi P, Nguyen JT, Gibbons CM, Veasey SC, Krueger JM. Tumor necrosis factor alpha in sleep regulation. Sleep Med Rev. 2018;40:69–78.
133. Probert L. TNF and its receptors in the CNS: The essential, the desirable and the deleterious effects. Neuroscience. 2015;302:2–22.
134. Dimitrov S, Besedovsky L, Born J, Lange T. Differential acute effects of sleep on spontaneous and stimulated production of tumor necrosis factor in men. Brain Behav Immun. 2015;47:201–10.
135. Bjurstrom MF, Olmstead R, Irwin MR. Reciprocal relationship between sleep macrostructure and evening and morning cellular inflammation in rheumatoid arthritis. Psychosom Med. 2017;79:24–33.
136. Dumaine JE, Ashley NT. Acute sleep fragmentation does not alter pro-inflammatory cytokine gene expression in brain or peripheral tissues of leptin-deficient mice. PeerJ. 2018;6:e4423.
137. Haack M, Pollmacher T, Mullington JM. Diurnal and sleep-wake dependent variations of soluble TNF- and IL-2 receptors in healthy volunteers. Brain Behav Immun. 2004;18:361–7.
138. Hakim F, Wang Y, Zhang SX, Zheng J, Yolcu ES, Carreras A, Khalyfa A, Shirwan H, Almendros I, Gozal D. Fragmented sleep accelerates tumor growth and progression through recruitment of tumor-associated macrophages and TLR4 signaling. Cancer Res. 2014;74:1329–37.
139. Bollinger T, Bollinger A, Skrum L, Dimitrov S, Lange T, Solbach W. Sleep-dependent activity of T cells and regulatory T cells. Clin Exp Immunol. 2009;155:231–8.
140. Horohov DW, Pourciau SS, Mistric L, Chapman A, Ryan DH. Increased dietary fat prevents sleep deprivation-induced immune suppression in rats. Comp Med. 2001;51:230–3.
141. Benca RM, Kushida CA, Everson CA, Kalski R, Bergmann BM, Rechtschaffen A. Sleep deprivation in the rat: VII. Immune function. Sleep. 1989;12:47–52.
142. Sa-Nunes A, Bizzarro B, Egydio F, Barros MS, Sesti-Costa R, Soares EM, Pina A, Russo M, Faccioli LH, Tufik S, Andersen ML. The dual effect of paradoxical sleep deprivation on murine immune functions. J Neuroimmunol. 2016;290:9–14.
143. Irwin M, Mascovich A, Gillin JC, Willoughby R, Pike J, Smith TL. Partial sleep deprivation reduces natural killer cell activity in humans. Psychosom Med. 1994;56:493–8.
144. De Lorenzo BH, de Oliveira Marchioro L, Greco CR, Suchecki D. Sleep-deprivation reduces NK cell number and function mediated by beta-adrenergic signalling. Psychoneuroendocrinology. 2015;57:134–43.
145. Hahn J, Gunter M, Schuhmacher J, Bieber K, Poschel S, Schutz M, Engelhardt B, Oster H, Sina C, Lange T, Autenrieth SE. Sleep enhances numbers and function of monocytes and improves bacterial infection outcome in mice. Brain Behav Immun. 2020;87:329–38.

146. Sang D, Lin K, Yang Y, Ran G, Li B, Chen C, Li Q, Ma Y, Lu L, Cui XY, Liu Z, Lv SQ, Luo M, Liu Q, Li Y, Zhang EE. Prolonged sleep deprivation induces a cytokine-storm-like syndrome in mammals. Cell. 2023;186:5500–16 e5521.
147. Hui L, Hua F, Diandong H, Hong Y. Effects of sleep and sleep deprivation on immunoglobulins and complement in humans. Brain Behav Immun. 2007;21:308–10.
148. Reis ES, Lange T, Kohl G, Herrmann A, Tschulakow AV, Naujoks J, Born J, Kohl J. Sleep and circadian rhythm regulate circulating complement factors and immunoregulatory properties of C5a. Brain Behav Immun. 2011;25:1416–26.
149. Zhai Q, Zhang Y, Ye M, Zhu S, Sun J, Wang Y, Deng B, Ma D, Wang Q. Reducing complement activation during sleep deprivation yields cognitive improvement by dexmedetomidine. Br J Anaesth. 2023;131:542–55.
150. Quan N. In-depth conversation: spectrum and kinetics of neuroimmune afferent pathways. Brain Behav Immun. 2014;40:1–8.
151. Dantzer R, Kelley KW. Twenty years of research on cytokine-induced sickness behavior. Brain Behav Immun. 2007;21:153–60.
152. Brown R, Price RJ, King MG, Husband AJ. Are antibiotic effects on sleep behavior in the rat due to modulation of gut bacteria? Physiol Behav. 1990;48:561–5.
153. Krueger JM, Pappenheimer JR, Karnovsky ML. The composition of sleep-promoting factor isolated from human urine. J Biol Chem. 1982;257:1664–9.
154. Archer AE, Von Schulze AT, Geiger PC. Exercise, heat shock proteins and insulin resistance. Philos Trans R Soc Lond B Biol Sci. 2018;373
155. Bierhaus A, Wolf J, Andrassy M, Rohleder N, Humpert PM, Petrov D, Ferstl R, von Eynatten M, Wendt T, Rudofsky G, Joswig M, Morcos M, Schwaninger M, McEwen B, Kirschbaum C, Nawroth PP. A mechanism converting psychosocial stress into mononuclear cell activation. Proc Natl Acad Sci U S A. 2003;100:1920–5.
156. Johnson JD, Campisi J, Sharkey CM, Kennedy SL, Nickerson M, Fleshner M. Adrenergic receptors mediate stress-induced elevations in extracellular Hsp72. J Appl Physiol (1985). 2005;99:1789–95.
157. Benedict C, Cedernaes J, Giedraitis V, Nilsson EK, Hogenkamp PS, Vagesjo E, Massena S, Pettersson U, Christoffersson G, Phillipson M, Broman JE, Lannfelt L, Zetterberg H, Schioth HB. Acute sleep deprivation increases serum levels of neuron-specific enolase (NSE) and S100 calcium binding protein B (S-100B) in healthy young men. Sleep. 2014;37:195–8.
158. Ooms S, Overeem S, Besse K, Rikkert MO, Verbeek M, Claassen JA. Effect of 1 night of total sleep deprivation on cerebrospinal fluid beta-amyloid 42 in healthy middle-aged men: a randomized clinical trial. JAMA Neurol. 2014;71:971–7.
159. Shokri-Kojori E, Wang GJ, Wiers CE, Demiral SB, Guo M, Kim SW, Lindgren E, Ramirez V, Zehra A, Freeman C, Miller G, Manza P, Srivastava T, De Santi S, Tomasi D, Benveniste H, Volkow ND. beta-Amyloid accumulation in the human brain after one night of sleep deprivation. Proc Natl Acad Sci U S A. 2018;115:4483–8.
160. Krueger JM, Opp MR. Sleep and microbes. Int Rev Neurobiol. 2016;131:207–25.
161. Krueger JM, Rector DM, Roy S, Van Dongen HP, Belenky G, Panksepp J. Sleep as a fundamental property of neuronal assemblies. Nat Rev Neurosci. 2008;9:910–9.
162. Krueger JM, Majde JA. Cytokines and sleep. Int Arch Allergy Immunol. 1995;106:97–100.
163. Opp MR. Cytokines and sleep. Sleep Med Rev. 2005;9:355–64.
164. Visan I. Mapping IL-1 in the brain. Nat Immunol. 2019;20:245.
165. Krumm B, Xiang Y, Deng J. Structural biology of the IL-1 superfamily: key cytokines in the regulation of immune and inflammatory responses. Protein Sci. 2014;23:526–38.
166. Liu T, Zhang L, Joo D, Sun SC. NF-kappaB signaling in inflammation. Signal Transduct Target Ther. 2017;2:17023.
167. Albensi BC. What is nuclear factor kappa B (NF-kappaB) doing in and to the mitochondrion? Front Cell Dev Biol. 2019;7:154.
168. Kumar H, Kawai T, Akira S. Pathogen recognition by the innate immune system. Int Rev Immunol. 2011;30:16–34.

169. Schroder K, Tschopp J. The inflammasomes. Cell. 2010;140:821–32.
170. Chen Z, Gardi J, Kushikata T, Fang J, Krueger JM. Nuclear factor-kappaB-like activity increases in murine cerebral cortex after sleep deprivation. Am J Physiol. 1999;276:R1812–8.
171. Kubota T, Fang J, Brown RA, Krueger JM. Interleukin-18 promotes sleep in rabbits and rats. Am J Physiol Regul Integr Comp Physiol. 2001;281:R828–38.
172. Takahashi S, Kapas L, Fang J, Krueger JM. An anti-tumor necrosis factor antibody suppresses sleep in rats and rabbits. Brain Res. 1995;690:241–4.
173. Fang J, Wang Y, Krueger JM. Mice lacking the TNF 55 kDa receptor fail to sleep more after TNFalpha treatment. J Neurosci. 1997;17:5949–55.
174. Miller AH, Raison CL. The role of inflammation in depression: from evolutionary imperative to modern treatment target. Nat Rev Immunol. 2016;16:22–34.
175. De A, Krueger JM, Simasko SM. Tumor necrosis factor alpha increases cytosolic calcium responses to AMPA and KCl in primary cultures of rat hippocampal neurons. Brain Res. 2003;981:133–42.
176. Furukawa K, Mattson MP. The transcription factor NF-kappaB mediates increases in calcium currents and decreases in NMDA- and AMPA/kainate-induced currents induced by tumor necrosis factor-alpha in hippocampal neurons. J Neurochem. 1998;70:1876–86.
177. Ye L, Huang Y, Zhao L, Li Y, Sun L, Zhou Y, Qian G, Zheng JC. IL-1beta and TNF-alpha induce neurotoxicity through glutamate production: a potential role for neuronal glutaminase. J Neurochem. 2013;125:897–908.
178. Zumkehr J, Rodriguez-Ortiz CJ, Medeiros R, Kitazawa M. Inflammatory cytokine, IL-1beta, regulates glial glutamate transporter via microRNA-181a in vitro. J Alzheimers Dis. 2018;63:965–75.
179. Tononi G, Cirelli C. Sleep and the price of plasticity: from synaptic and cellular homeostasis to memory consolidation and integration. Neuron. 2014;81:12–34.
180. Lai AY, Swayze RD, El-Husseini A, Song C. Interleukin-1 beta modulates AMPA receptor expression and phosphorylation in hippocampal neurons. J Neuroimmunol. 2006;175:97–106.
181. Ricciotti E, FitzGerald GA. Prostaglandins and inflammation. Arterioscler Thromb Vasc Biol. 2011;31:986–1000.
182. Vane JR. Inhibition of prostaglandin synthesis as a mechanism of action for aspirin-like drugs. Nat New Biol. 1971;231:232–5.
183. Inoue S, Honda K, Komoda Y, Uchizono K, Ueno R, Hayaishi O. Differential sleep-promoting effects of five sleep substances nocturnally infused in unrestrained rats. Proc Natl Acad Sci U S A. 1984;81:6240–4.
184. Matsumura H, Nakajima T, Osaka T, Satoh S, Kawase K, Kubo E, Kantha SS, Kasahara K, Hayaishi O. Prostaglandin D2-sensitive, sleep-promoting zone defined in the ventral surface of the rostral basal forebrain. Proc Natl Acad Sci U S A. 1994;91:11998–2002.
185. Terao A, Matsumura H, Yoneda H, Saito M. Enhancement of slow-wave sleep by tumor necrosis factor-alpha is mediated by cyclooxygenase-2 in rats. Neuroreport. 1998;9:3791–6.
186. Yoshida H, Kubota T, Krueger JM. A cyclooxygenase-2 inhibitor attenuates spontaneous and TNF-alpha-induced non-rapid eye movement sleep in rabbits. Am J Physiol Regul Integr Comp Physiol. 2003;285:R99–109.
187. Ram A, Pandey HP, Matsumura H, Kasahara-Orita K, Nakajima T, Takahata R, Satoh S, Terao A, Hayaishi O. CSF levels of prostaglandins, especially the level of prostaglandin D2, are correlated with increasing propensity towards sleep in rats. Brain Res. 1997;751:81–9.
188. Steptoe A, Hamer M, Chida Y. The effects of acute psychological stress on circulating inflammatory factors in humans: a review and meta-analysis. Brain Behav Immun. 2007;21:901–12.
189. Besedovsky H, del Rey A, Sorkin E, Dinarello CA. Immunoregulatory feedback between interleukin-1 and glucocorticoid hormones. Science. 1986;233:652–4.
190. Dantzer R, O'Connor JC, Freund GG, Johnson RW, Kelley KW. From inflammation to sickness and depression: when the immune system subjugates the brain. Nat Rev Neurosci. 2008;9:46–56.
191. Blalock JE. The immune system as the sixth sense. J Intern Med. 2005;257:126–38.

192. Blalock JE, Smith EM. Conceptual development of the immune system as a sixth sense. Brain Behav Immun. 2007;21:23–33.
193. McCusker RH, Kelley KW. Immune-neural connections: how the immune system's response to infectious agents influences behavior. J Exp Biol. 2013;216:84–98.
194. Dimitrov S, Lange T, Nohroudi K, Born J. Number and function of circulating human antigen presenting cells regulated by sleep. Sleep. 2007;30:401–11.
195. Flierl MA, Rittirsch D, Nadeau BA, Chen AJ, Sarma JV, Zetoune FS, McGuire SR, List RP, Day DE, Hoesel LM, Gao H, Van Rooijen N, Huber-Lang MS, Neubig RR, Ward PA. Phagocyte-derived catecholamines enhance acute inflammatory injury. Nature. 2007;449:721–5.
196. Ordovas-Montanes J, Rakoff-Nahoum S, Huang S, Riol-Blanco L, Barreiro O, von Andrian UH. The regulation of immunological processes by peripheral neurons in homeostasis and disease. Trends Immunol. 2015;36:578–604.
197. Pert CB, Ruff MR, Weber RJ, Herkenham M. Neuropeptides and their receptors: a psychosomatic network. J Immunol. 1985;135:820s–6s.
198. Chavan SS, Pavlov VA, Tracey KJ. Mechanisms and therapeutic relevance of neuro-immune communication. Immunity. 2017;46:927–42.
199. Fujii T, Mashimo M, Moriwaki Y, Misawa H, Ono S, Horiguchi K, Kawashima K. Physiological functions of the cholinergic system in immune cells. J Pharmacol Sci. 2017;134:1–21.
200. Berczi I, Quintanar-Stephano A, Kovacs K. Neuroimmune regulation in immunocompetence, acute illness, and healing. Ann N Y Acad Sci. 2009;1153:220–39.
201. Dorshkind K, Horseman ND. Anterior pituitary hormones, stress, and immune system homeostasis. Bioessays. 2001;23:288–94.
202. Nezi M, Mastorakos G, Mouslech Z. Corticotropin releasing hormone and the immune/inflammatory response. In: Feingold KR, Anawalt B, Blackman MR, Boyce A, Chrousos G, Corpas E, de Herder WW, Dhatariya K, Dungan K, Hofland J, Kalra S, Kaltsas G, Kapoor N, Koch C, Kopp P, Korbonits M, Kovacs CS, Kuohung W, Laferrere B, Levy M, McGee EA, McLachlan R, New M, Purnell J, Sahay R, Shah AS, Singer F, Sperling MA, Stratakis CA, Trence DL, Wilson DP, editors. Endotext. South Dartmouth, MA; 2000.
203. Weigent DA. Lymphocyte GH-axis hormones in immunity. Cell Immunol. 2013;285:118–32.
204. Gross PM. Circumventricular organ capillaries. Prog Brain Res. 1992;91:219–33.
205. Miyata S. New aspects in fenestrated capillary and tissue dynamics in the sensory circumventricular organs of adult brains. Front Neurosci. 2015;9:390.
206. Boulanger LM. Immune proteins in brain development and synaptic plasticity. Neuron. 2009;64:93–109.
207. Serrats J, Schiltz JC, Garcia-Bueno B, van Rooijen N, Reyes TM, Sawchenko PE. Dual roles for perivascular macrophages in immune-to-brain signaling. Neuron. 2010;65:94–106.
208. Steinman L. Elaborate interactions between the immune and nervous systems. Nat Immunol. 2004;5:575–81.
209. Wohleb ES, McKim DB, Sheridan JF, Godbout JP. Monocyte trafficking to the brain with stress and inflammation: a novel axis of immune-to-brain communication that influences mood and behavior. Front Neurosci. 2014;8:447.
210. Sandiego CM, Gallezot JD, Pittman B, Nabulsi N, Lim K, Lin SF, Matuskey D, Lee JY, O'Connor KC, Huang Y, Carson RE, Hannestad J, Cosgrove KP. Imaging robust microglial activation after lipopolysaccharide administration in humans with PET. Proc Natl Acad Sci U S A. 2015;112:12468–73.
211. Schedlowski M, Engler H, Grigoleit JS. Endotoxin-induced experimental systemic inflammation in humans: a model to disentangle immune-to-brain communication. Brain Behav Immun. 2014;35:1–8.
212. Bordoni B, Purgol S, Bizzarri A, Modica M, Morabito B. The influence of breathing on the central nervous system. Cureus. 2018;10:e2724.
213. Garcia-Medina NE, Miranda MI. Nucleus of the solitary tract chemical stimulation induces extracellular norepinephrine release in the lateral and basolateral amygdala. Brain Stimul. 2013;6:198–201.

214. Pavlov VA, Tracey KJ. The vagus nerve and the inflammatory reflex—linking immunity and metabolism. Nat Rev Endocrinol. 2012;8:743–54.
215. Hansen MK, Krueger JM. Subdiaphragmatic vagotomy blocks the sleep- and fever-promoting effects of interleukin-1beta. Am J Physiol. 1997;273:R1246–53.
216. Kubota T, Fang J, Guan Z, Brown RA, Krueger JM. Vagotomy attenuates tumor necrosis factor-alpha-induced sleep and EEG delta-activity in rats. Am J Physiol Regul Integr Comp Physiol. 2001;280:R1213–20.
217. Zielinski MR, Dunbrasky DL, Taishi P, Souza G, Krueger JM. Vagotomy attenuates brain cytokines and sleep induced by peripherally administered tumor necrosis factor-alpha and lipopolysaccharide in mice. Sleep. 2013;36:1227–38. 1238A
218. Hansen MK, Taishi P, Chen Z, Krueger JM. Vagotomy blocks the induction of interleukin-1beta (IL-1beta) mRNA in the brain of rats in response to systemic IL-1beta. J Neurosci. 1998;18:2247–53.
219. Laye S, Bluthe RM, Kent S, Combe C, Medina C, Parnet P, Kelley K, Dantzer R. Subdiaphragmatic vagotomy blocks induction of IL-1 beta mRNA in mice brain in response to peripheral LPS. Am J Physiol. 1995;268:R1327–31.
220. Kaur C, Ling EA. The circumventricular organs. Histol Histopathol. 2017;32:879–92.
221. Xanthos DN, Sandkuhler J. Neurogenic neuroinflammation: inflammatory CNS reactions in response to neuronal activity. Nat Rev Neurosci. 2014;15:43–53.
222. Bourhy L, Mazeraud A, Bozza FA, Turc G, Lledo PM, Sharshar T. Neuro-inflammatory response and brain-peripheral crosstalk in sepsis and stroke. Front Immunol. 2022;13:834649.

Sleep and Circadian Disruption in Depression and Anxiety Disorders: An Inflammatory Insight

Meena Mirdha and Lal Chandra Vishwakarma

Abstract Sleep and circadian rhythm are reciprocally interlinked with mental health. Disturbed sleep or insufficient sleep activates and propagates inflammatory pathways. Sleep deprivation alone can stimulate the toll-like receptor (TLR) pathway and induce cytokine production. Inflammatory markers act under circadian rhythms and fluctuations in blood levels are seen; interestingly, these fluctuations depend on a particular time of the day. In addition to immunosuppression, major depression relates to inflammatory activation. Major depressive disorder (MDD) is distinguished by the activation of T cell with increased T helper (Th)-1 product levels such as serum IL-2 and soluble IL-2 receptor (sIL-2R) in serum. Coexisting anxiety and depression are a more severe disorder than depression alone. Studies showed that hypothalamic-pituitary-adrenal (HPA) axis dysregulation is associated with late-life post-traumatic stress disorder (PTSD). We aim to integrate sleep and circadian disruption with inflammatory pathways in depression and anxiety disorders.

Keywords Anxiety · Depression · Circadian rhythm · Inflammatory pathways Hypothalamic-pituitary-adrenal axis

1 Introduction

Inflammation is generally an effort by immune cells to annihilate the earliest cause of cell injury. Human and animal studies have shown that specific immune and inflammatory molecules and mechanisms alter the circadian clock and regulate

M. Mirdha (✉)
Department of Physiology, All India Institute of Medical Sciences, Bathinda, Punjab, India

L. C. Vishwakarma
Division of Sleep Medicine, Department of Medicine, University of Pennsylvania, Philadelphia, PA, USA

P. K. Jha, H. Bouaouda (eds.), *Circadian Rhythms, Sleep and Inflammation*, Progress in Inflammation Research 93,
https://doi.org/10.1007/978-3-031-74744-1_6

sleep [1–4]. Inflammation allows survival during infection and injury and plays a role in tissue homeostasis. The inflammatory process comes with a brief decline in tissue function. Recently, studies observed the activation and spreading of inflammatory pathways in disturbed sleep or where sleep is insufficient. Rechtschaffen et al. evidenced that sleep deprivation inflicts significant undesirable consequences on the ability to overcome infection [5]. The immune system and sleep relationship are reciprocal. The production of inflammatory cytokines affects sleep. Cytokines as immunomodulators have been extensively studied in terms of the functions they play and result in host responses to infection. It has a role as a regulator of physiological sleep. The acute phase response indicates prompt actuation of the immune system in retaliation to infection injury. The innate immune system's components constitute toll-like receptors (TLRs), which are pattern recognition receptors that are conscious of self and non-self-elements that are demonstrated on tissue macrophages which instigate the making of cytokines, such as tumor necrosis factor (TNF-α), interleukin (IL-1), IL-6 and chemokines (e.g. CCL2, CXCL8), and prostaglandins [6, 7]. Outer lipopolysaccharides (LPS) of bacterial cell membranes classically activate these pathways [8]. Sleep deprivation can stimulate the toll-like receptor pathway and produce cytokine. Irwin et al. observed that after a disturbed night in which loss of sleep has occurred, in that case, activation of TLR-4 was higher than that after undisturbed sleep [9]. Nuclear factor (NF-kB) is a transcription factor cardinal in activation of pathways of inflammation and is associated with the transcription of many genes, which are inclusive of the making of cytokines and inflammatory markers [10, 11]. NF-kB normally exists in the cytosol in a bound form, which makes it inactive. NF-kB binding areas are set up in the promoter regions of the genes of sleep-inducing matter, like adenosine A1 receptors, cyclooxygenase-2 (Cox-2), and nitric oxide synthase (NOS) [12]. Therefore, NF-kB is a critical factor in activating and regulating pathways of the inflammatory process, associated with gene transcription that is accountable for producing inflammatory markers and cytokines [11]. Even mild sleep deprivation can influence the actuation of NF-kB and its translocation in particular brain areas interrelated with the regulation of sleep [13–16]. Sleep deprivation alone can stimulate the TLR pathway and induce cytokine production, leading to local inflammatory responses [17]. Cytokines and chemokines vastly studied in animals as well as in humans and evidenced to influence sleep include IL-1α, IL-1 β, IL-2, IL-4, IL-6, IL-8, IL-10, IL-13, IL-15, IL-18, TNF-α, TNF-β, IFN-α, IFN-β, and macrophage inhibitory protein [18, 19]. The majority of cytokines were earlier observed in the immune system in the periphery; many cytokines and respective receptors are observed to be present in the central nervous system (CNS) and localized to specific brain regions in current times [20, 21]. One study showed injection of IL-1 and TNF-α in animals lengthen non-rapid eye movement (NREM) sleep time. Additionally, when these substances are blocked, they disturb spontaneous sleep as well as inhibit sleep rebound after sleep deprivation [22]. These two cytokines reveal mutually dependent results on sleep, as observed by the use of TNF receptor (TNFRF) and IL-1 receptor fragment (IL-1RF) [23]. In a study, human subjects were given a parenteral dose of IL-6 adequate to attain systemic infection. Following six and half hours, IL-6 brought out subjective fatigue and led to an increment in C-reactive protein (CRP) in contrast to placebo. Suppression of rapid eye movement (REM) sleep was also seen while slow-wave sleep (SWS) time showed akin to the placebo. SWS time decreased during the first half of the sleep period and increased during the last half of sleep [24]. Surprisingly, other inflammatory cytokines levels after injection of IL-6 showed no increase, put forward that IL-6 is notably pertinent to variations in patterns of sleep [25].

So, on a large scale, IL-1, IL-6, and TNF are studied in human subjects and laboratory animals [26]. It depends on the particular time of the day; particular inflammatory markers act under circadian rhythms, and their fluctuations in blood levels are seen. Surprisingly, a circadian profile of systemic levels of IL-6 peaks at times 7 pm and 5 am. Night sleep is requisite for a nighttime increase in IL-6 levels and for making TNF by stimulated monocytes [27]. Sleep deprivation affects IL-6 secretion from night to day and shifts their pattern, leading to IL-6 over-secretion during the day [28]. Adults presenting with short sleep duration in objective assessment and insomnia are at high risk of neurocognitive impairment, depression, hypertension, and mortality [29]. Studies showed that stress activates inflammatory pathways and suppresses the system of acquired immunity. Recently, it has been evident that in addition to immunosuppression, major depression relates to inflammatory activation [30]. It has been proposed that chronic low-grade inflammation is one of the ways through which insomnia can commence to altered health consequences. Short sleep duration and insomnia in adolescents are also related to systemic inflammation [31]. Fernandez-Mendoza et al. studied 378 adolescents and documented insomnia or shorter sleep duration in a polysomnographic study. A fasting blood was drawn to estimate CRP plasma levels and other inflammatory markers. They observed elevated CRP levels in adolescent subjects, who also reported insomnia symptoms and slept for 6–7 h objectively. This interrelation does not depend on demographic factors associated with insomnia symptoms, evening circadian preference, or any medical condition [29].

2 Circadian Rhythms in Psychiatric Disorders

Depression is quite common. The major depressive disorder lifetime prevalence of adults in the United States is 20.6%, whereas the annual prevalence is 10.4%. Men are less commonly affected than women [30]. Subjects with depression make alterations in circadian rhythm and diurnal mood variations, which also lead to sleep disturbances. Chronotherapies like bright light therapies, sleep deprivation, and social rhythm therapies are found as helpful supplements in seasonal and non-seasonal depression [31].

Recently, a study investigated the correlation of diurnal mood variation in patients who are in depression and compared them with healthy subjects and observed that subjects with depression show discrete variations of brain glucose metabolism regionally over times of the day in comparison with participants who are not in depression. Additionally, improvement in mood occurs in the evening in subjects who are depressed, which is linked with the dorsal neural network activation [32]. In patients with depression, it is observed that activity sustained in the hypothalamic regions and brainstem is associated with wakefulness throughout the daytime. In contrast, healthy individuals have increased brain glucose metabolism in the evening [33]. Recently, researchers showed that oscillations in plasma cortisol and norepinephrine are phase-advanced in subjects with depression in comparison with normal individuals [34]. In mood-disordered subjects, subjective sleep complaints are commonly seen. Around 90% of subjects with depression witnessed difficulty falling asleep, difficulty remaining asleep, and early morning awakenings [35, 36].

2.1 Major Depressive Disorder with Seasonal Pattern

This seasonal form of depression with characteristics equivalent to major depressive disorder (MDD) occurring on a cyclical basis is linked to light deprivation during the winter season. Light therapy and medications are more often practiced as treatment. Winter-onset *seasonal affective disorder* (SAD) is linked with hypersomnia. In contrast, summer-onset SAD is linked to insomnia [37]. The phase-shift conjecture of depression puts forward disturbances in mood leading from a delay of the central pacemaker and related circadian rhythms that take part in the regulation of temperature, cortisol, melatonin, and REM sleep concerning other circadian rhythms. The final result showed that the advanced circadian phase, like early morning awakenings, early occurrence of REM sleep, and melatonin secretion, shifts in individuals with depression compared to non-depressed individuals. Treatment for phase shift involves bright light exposure and melatonin to rearrange the internal rhythms and the sleep–wake cycle. These showed effective results, especially in patients with winter depression [38–41].

2.2 Non-seasonal Unipolar Depression and Circadian Rhythms

In non-seasonal unipolar depression, abnormal circadian rhythms are observed. Individuals with depression are more likely to be evening chronotypes and show delayed rhythms and reduced peak amplitude of melatonin and core body temperature compared to healthy individuals. Variants in clock genes like CLOCK and CRY are associated with depression diagnosis. In these subjects, sleep disturbance, specifically insomnia, is prevalent. Moreover, REM latency, or from sleep onset time to the first REM episode, is reduced, but the duration of REM is lengthened in depression. A recent curious concept about sleep and depression is that total sleep deprivation can offer a transient antidepressant effect in most patients, which contributes to approximately 60% of depressed individuals. This wake therapy includes sleep restriction for a night or the latter half of the night. Still, it cannot be used as a monotherapy as symptoms of depression will return after a full night of sleep, and the treatment effect needs to be maintained with another type of treatment. Other chronotherapies include advancing sleep phase, bright light therapy, and melatonin administration. Wake therapy can result in fast antidepressant effects. Hence, the combination of wake therapy with antidepressant medications and/or light therapy shows a promising response followed by sufficient maintenance of the antidepressant response. In addition, it shows that effective antidepressant medications have therapeutic outcomes on sleep processes, like inhibiting excessive REM sleep or lengthening REM latency, which conceivably reflects underlying circadian resynchronization. Agomelatine, an antidepressant medication that acts as a melatonin receptor agonist, increases monoaminergic activity. Melatonin or agomelatine administration 5 h before bedtime can advance rhythms and improve mood. Circadian regulation of hormonal, temperature, and mood rhythms differs; social *zeitgebers* and genetic factors can influence it and are likely to interact, underlying the complexity of circadian influences on mood disorders [42]. In a study by Deltito et al. conducted on 17 patients of non-seasonal depressive disorder, there was a comparison of the reaction of bipolar spectrum versus unipolar spectrum patients to the treatment of light therapy. They treated all patients with 400 or 2500 lux phototherapy for two hours on seven successive days and observed improvement in all bipolar versus unipolar spectrum patients. All patients who responded to phototherapy were observed for 3 months, and they maintained their effects at a 3-month follow-up [43].

In patients with seasonal patterns of mild and moderate depression, light therapy is proposed as the first-line treatment. For non-seasonal mild and moderate depression, light therapy is used as an additional treatment.

2.3 Neurobiology of Depressive Disorders

Depression is linked with an untimely loss of slow-wave sleep and more arousal. Disturbance types are set back by nocturnal arousal that is listed as (1) increased nighttime arousals, (2) a decrease in total sleep time, (3) phasic REM sleep increased, and (4) core body temperature increased. An amalgamation of increased REM drive and reduced slow-wave sleep results in a decrease in the first period of NREM sleep; this refers to a decrease in REM latency. Twin studies, including family studies, showed that abnormalities are partially heritable. Inheritable trait behavior shows reduced REM latency, and slow-wave sleep deficits remain after recovering from depressive episodes. Difficulties in maintaining sleep and increased phasic REM sleep were linked to increased cortisol, increased limbic blood flow, and glucose metabolism. The anti-histaminic and antagonistic effects of 5-HT2 and melatonin receptors primarily mediate the effects of the antidepressants on sleep maintenance and slow-wave sleep. Sleep abnormality has been examined in depressed youth. Some studies showed rapid eye movement latencies are shorter, while sleep latencies are longer, REM periods are in high number, and REM time increased for children who were in depression. Studies in adolescents who are depressed showed REM latency is short, REM sleep increased, REM density increased, and long sleep latency. Reports from youth of subjective sleep disturbances (sleep quality, waking in mid-night, tiredness in the morning) are not confirmed by electroencephalogram (EEG) or objective examination. Research on the recurrence of depression in youth examined using polysomnography showed those youth who had a later recurrence of depression also had reduced sleep efficiency and sleep onset delayed at baseline polysomnography [42].

MDD is distinguished by the activation of T cells with increased T helper (Th)-1 product levels like serum IL-2, soluble IL-2 receptor (sIL-2R) in serum, and formation of interferon (IFN)-γ [44]. Activation of T cells in MDD give a demonstration of increase expression of CD4+ , CD4+/CD7 + CD25+ and CD4 + HLA-DR + cell surface markers in examination with flow cytometry [45–47]. Various research findings showed that MDD is linked with increased M1 macrophage activation, observed as elevated polyclonally stimulated production of IL-1β , IL-6, elevated serum IL-6, and tumor necrosis factor levels [48–51]. In MDD, complement factor levels increased and lowered negative acute phase protein levels indicate an existing mild inflammatory reaction [52, 53]. Immune-inflammatory response system (IRS) (activated M1, Th-1, Th-17) and compensatory immunoregulatory system (CIRS) (Th-2, Treg) are activated in MDD [44]. Researchers found that TNF-α, IL-6, CRP, IL-1β, IL-2, and sIL-1RA were elevated in drug-naïve first episode MDD significantly [54, 55]. However, IL-8 was significantly decreased [54].

According to findings from recent research, MDD is categorized into two types: simple dysmood disorder (SDMD) and major dysmood disorder (MDMD) [56]. Stimulated production of Th-1, Th-2, Th-17, M1, Treg, IRS, CIRS, and growth factors are comparatively increased in MDMD than in control subjects. Interestingly, some M1(IL-1β, IL-15, IL-6, CCL2, CCL5, CXCL10, TNF-α), Th1(IL-2, IFN-γ, IL-16), and Th-17(IL-6 and IL-17) even Th-4 (IL-4) products show an effect like neurotoxicity so, altered neuronal functions of regulating affection and lead to neurotoxicity and depression. So, MDMD is a subtype of MDD which is inflammation-associated [57].

2.4 Anxiety

Neurochemicals are usually released during stress or in periods of anxiety, which leads to alteration in the neurobiological system. Severe stress causes maladaptive results when it affects biological systems to be chronically activated. Patients with all types of anxiety disorders show symptoms consistent with an exaggerated noradrenergic system output. Serotonin (5-HT) is secreted in response to stress. In younger adults, comorbidity of anxiety and mood disorders is a common observation. In a recent research study, it was observed that 21–37% of patients with an anxiety disorder also experienced comorbid depression. Among nursing home patients, 42% of patients with anxiety disorder experienced comorbid major depression. However, this overlap is not found in older adults. Older age appears to be protective for anxiety but a risk factor for depression. Coexisting anxiety and depression are a more severe disorder than depression alone. Studies showed that hypothalamic pituitary adrenal axis dysregulation is associated with late-life post-traumatic stress disorder (PTSD). Patients with PTSD show flatter circadian salivary cortisol patterns and lower urinary cortisol excretion than those without PTSD [1]. In PTSD subjects, atrophy of the hippocampus has been delineated, indicating that it is a ubiquitous task considered chronic dysbalance in the adaptation chronically. This dysbalance comprises the HPA axis and endogenous neurotransmitters [58]. In anxiety disorders, sleep disturbances are common; sometimes, this is the main complaint by most patients. These patients present sleep problems as difficulty in the initiation and maintenance of sleep, and nightmares occur as sleep disturbances in these patients. Arousals occur additionally from NREM sleep along with the apparent nature of REM-related disturbances. An integrated response to stress, mainly emotional stress, is shown in Fig. 1. Stress signals evoke combined retaliation of sympatho-adrenal-medullary system and the HPA axis. Steps involved in stress adaptation by multiple physiological levels are shown. Stress is discerned by brain, from the cortex to the brainstem, which is inclusive of the reticular activating system (RAS) and limbic system. Stress stimulates hypothalamus and activates the corticotrophin-releasing hormone (CRH) and antidiuretic hormone (ADH) neurons in the paraventricular nucleus and also in adrenergic neurons. ADH release stimulates adrenocorticotrophic hormone (ACTH) release and increases cortisol release. The combined role of the HPA axis and sympathoadrenal medullary system in stress adaptation is a necessary physiological phenomenon. This is maintained by physiological actions, such as an increment in glucose production (catecholamines increase plasma glucose by glycogenolysis activation, and cortisol action done by giving amino acids substrate for gluconeogenesis. They switch glucose utilization on the way to the central nervous system compared to the peripheral tissues. Epinephrine increases the free fatty acids supply to the heart and muscles, and cortisol enhances lipolysis. Catecholamines and cortisol increase cardiac output, improving the substrate supply to tissues. Arousal results from adrenergic stimuli to brain centers. It also interacts to immune system to make a balance between local cytokine production. Anti-inflammatory effects are produced by inhibiting NF-kB activation. In anxiety disorders, the HPA axis dysbalance is encountered.

A study of brain function in depressed patients during sleep showed sleep disturbances like hypersomnia, difficulty maintaining sleep, and increased arousal in depressive patients [29]. Studies also showed cerebral metabolism at the time of NREM sleep in depressed subjects is higher in comparison with that in normal subjects. Following 24 h of total sleep deprivation in normal subjects, it was found that cerebral glucose metabolism decreased by 8%. It is prominent in the frontal and

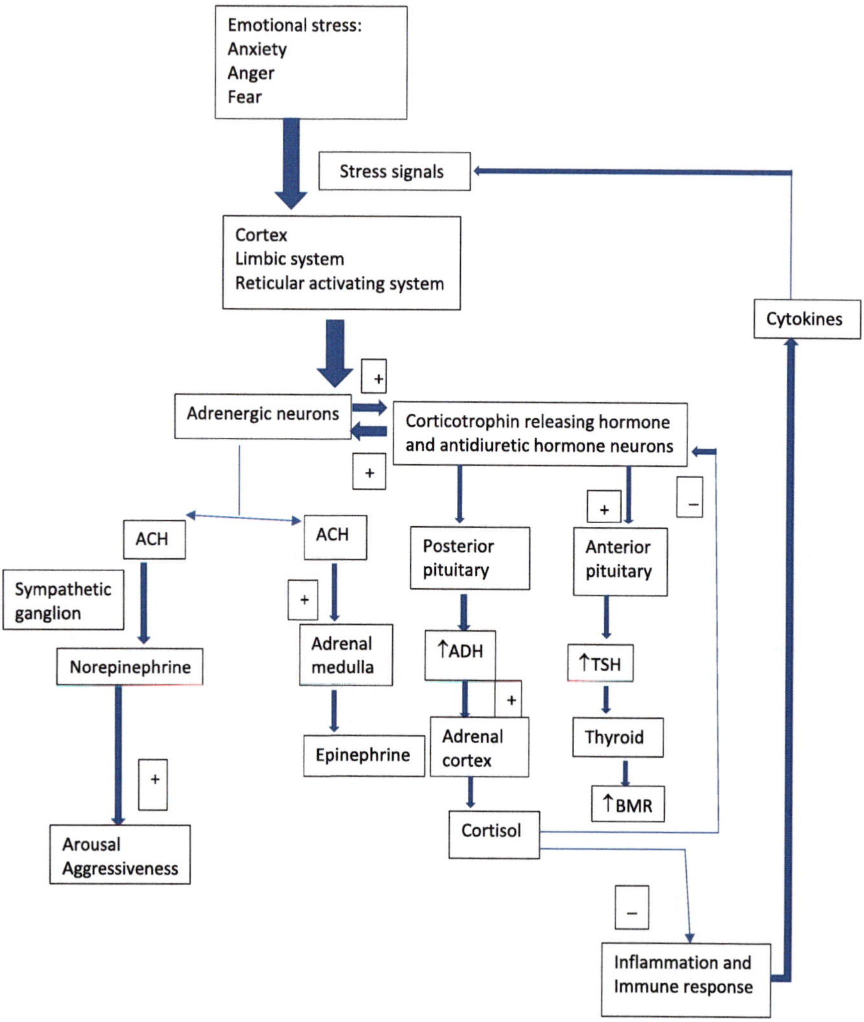

Fig. 1 Steps involved in the adaptation to stress by an integrated response of hypothalamic-pituitary-adrenal axis and sympathoadrenal medullary system

parietal lobes' thalamic nuclei and associative cortices. In subjects with depression who respond to sleep deprivation, observed in awake patients' baseline brain activity is higher in the cingulate cortex or nearby mesial frontal cortex when compared with non-responders. These changes normalize after sleep deprivation. A similar pattern is also found in older subjects who are depressed. This study shows the close connection between mood alteration and paralimbic and limbic structure activity. In depressed patients during wakefulness, the anterior cingulate cortex was hyperactive. Sleep deprivation decreases depression ratings and activity in the anterior cingulate cortex [59].

3 Biological Dysfunction

3.1 The Inflammation Hypothesis

It represents that aging-related immune processes lead the way to changes in emotional and cognitive neural networks, making them susceptible to geriatric depression and activating metabolic brain changes leading to late-life depressive syndrome. Findings suggest that (1) aging leads to a proinflammatory state; (2) immune responses of the CNS have an impact upon the functionality of emotional and cognitive networks relevant to depression in older patients; (3) elevation of peripheral cytokines is associated with depressive symptoms in geriatric population; (4) Few antidepressants lower expression of inflammation markers in the periphery; and (5) plasma levels of cytokine before treatment may affect response to antidepressants.

Aging contributes to the chronic state of neuroinflammation. Studies on animals and humans showed that this proinflammatory state might compromise neural systems related to depression. In normal adults, subgenual anterior cingulate cortex activity increased, and, functional connectivity (FC) of the subgenual anterior cingulate with amygdala decreased, medial frontal cortex and nucleus accumbens in the course of emotional face processing were regulated by a peripheral inflammatory response (IL-6). A single nucleotide polymorphism (SNP) encoding IL-1β was observed to be as connected with both decreased activity of the anterior cingulate and the amygdala in rejoinder to emotional probes and with a lack of remark to antidepressants. In suicide victim's brain autopsies, it was found that in the orbitofrontal cortex, there are increased cytokine concentrations [42]. In a 6-year follow-up of older adults, elevation of IL-1Ra was a determinant for developing depressive symptoms [60]. Clinical studies of depressed patients documented a plasma elevation of many inflammatory markers (IL-6, TNF-α and IL-1β) and a correlation between severity of depressed mood and correlation with cognitive symptoms of depression. Tricyclic antidepressants (TCAs) may decrease some proinflammatory cytokines, increase anti-inflammatory cytokines, and reduce polysaccharide-induced "sickness behavior" in animals. Additionally, plasma levels of cytokine before treatment may control the response to antidepressants in humans. A higher level of TNF-α was shown to predict failure of depression to respond to escitalopram. A recent study showed that subjects with baseline low C-reactive protein (CRP < 1 mg/L) levels had greater improvement in depression when treated with escitalopram than with nortriptyline, while patients with high CRP levels had greater improvement when treated with nortriptyline than with escitalopram. Major depression, which is resistant to treatment, shows baseline concentrations of high-sensitivity CRP (>5 mg/L), TNF, and TNF-soluble receptors were associated with favorable response to the TNF antagonist infliximab [42]. Epidemiological studies have shown an association between the elevation of peripheral cytokines and depressive symptoms in older age groups. Increased IL-6 was the most accordant finding [61].

One mouse model study of chronic environmental circadian desynchronization (ECD) gave cognizance to limited literature. This new model is influenced by keeping mice to a light-dark cycle of twenty hours, which comprises ten hours of light and ten hours of dark. While other studies use standard laboratory 24-h cycles. So, this new model observed that after 5 weeks of environmental disruption, the animals started to reveal metabolic indications of cumulative allostatic load, increased adiposity, leptin levels, and weight gain. Additionally, it showed a disproportion between insulin and plasma glucose. Prefrontal cortex cellular morphology changes mimics that are perceived as stress. The ECD mice showed shrunken and less complex apical dendritic trees of cells in layer II/III of the medial prefrontal cortex, while on basal dendrites, there was no effect. It also shows abnormalities in behavior with prominent cognitive rigidity in a modified Morris water maze. The ECD animals showed standard performance on the preliminary task but were slow to adapt to a hidden platform moved to a new location. This type of variation is due to the mice making more errors by coming back to the original location of the platform [62]. For support of cognitive flexibility problems, ECD animals showed an impulsive phenotype in the light-dark box but did not show any overt anxiety-like phenotype. This effect of ECD is analogous to those studies that observed 21 days of moderate stress in rodents, producing morphologically simplifying prefrontal cortical neurons and dysbalance in prefrontal-mediated behaviors like attentional set-shifting or another task of working memory. The mechanism by which deranged circadian clocks lead to behavioral changes in brain and periphery is not very clear. However, one conjecture is that disturbed light-dark cycles open a progressive loss of continuity between circadian clocks in the brain (suprachiasmatic nucleus) and in the body (liver). The loss leads to internal asynchrony. This leads to altered synchrony between neural circuits in the brain (prefrontal cortex and amygdala). For long time, chronic circadian disruption could change these circuits, make them more vulnerable to subsequent insult, and could be a surreptitious mechanism of circuit-level disturbance, setting the environment for another stressor to an already risked network. Disturbance of the peripheral body clock changes how the stress system responds to stressors. This type of model embellishes in what form circadian disturbance could induce neural circuits to become more vulnerable to insults. So, these circadian disruption effects mimic the diathesis-stress models and correlate observations of increased risk for the development of psychiatric, and other physiologic dysbalance in the population facing circadian disruption [58].

3.2 *Circadian Biology to Advance Therapeutics*

Circadian disruptions are interlaced with mood disorders. So, pharmacological intervention of circadian dysbalance helps treat mood disorders. Intervention pharmacologically is the key component of the molecular circadian system that harmonizes circadian period, phase, and amplitude. It targets enzymes responsible for the phosphorylation of the clock proteins. Therefore, the molecules that target clock-regulatory kinases can change the circadian parameters of core clock genes and clock-controlled genes (CCGs) [63, 64]. The selective serotonin reuptake inhibitors

(SSRIs), popular antidepressants, reduce the circadian period by inhibiting the serotonin transporter activity [65]. Lithium is the mood stabilizer, glycogen synthase kinase (GSK3β) inhibitor that increases the circadian amplitude and period, which is demonstrated by measuring clock gene expression rhythms in mice [66, 67].

Slow-release formulations of melatonin and agomelatine show anxiolytic properties [68, 69]. A melatonin receptor agonist, ramelteon, is used in insomnia subjects comorbid with depression. This normalizes the expression of circadian genes like circadian locomotor output cycles kaput (CLOCK), cryptochrome (CRY1), CRY2, nuclear receptor subfamily (NR1D1), NR1D2, timeless circadian regulator (TIMELESS) [70].

4 Comorbid Medical and Psychiatric Illness

Subjects with MMD are at high risk of getting further comorbid disorders; common among them are panic disorder, alcohol dependence, social anxiety disorder, and obsessive-compulsive disorders. In both unipolar and bipolar disorder, men more often present with substance use disorder, while women often present with anxiety and eating disorders. In the National Sleep Foundation Survey 2003, adults aged 65 years and older, those with more medical conditions reported significantly more sleep complaints. Psychiatric disorders, such as mood disorders and anxiety, are also commonly associated with insomnia. In the survey, insomnia was present in 65% of respondents with major depression, 61% of those having panic disorder, and 44% of subjects with generalized anxiety disorder. World Mental Health (WMH) surveys provided lifetime and 12-month prevalence rates estimated for eighteen countries and divided into high and low-middle-income countries. The lifetime prevalence estimates are an average of 11.1 (range 8.0–18.4) in low and 14.6 (range 6.6–21.0) in high-income countries [42].

Compliance with Ethical Standards This chapter does not contain any studies involving animals and human participants performed by any of the authors.

Funding The authors declare no conflict of interest.

References

1. Krueger JM, Frank MG, Wisor JP, Roy S. Sleep function: toward elucidating an enigma. Sleep Med Rev. 2016;28:46–54.
2. Zielinski MR, Gerashchenko D, Karpova SA, Konanki V, McCarley RW, Sutterwala FS, et al. The NLRP3 inflammasome modulates sleep and NREM sleep delta power induced by spontaneous wakefulness, sleep deprivation and lipopolysaccharide. Brain Behav Immun. 2017;62:137–50.

3. Zielinski MR, McKenna JT, McCarley RW. Functions and mechanisms of sleep. AIMS Neurosci. 2016;3(1):67–104.
4. Besedovsky L, Lange T, Haack M. The sleep-immune crosstalk in health and disease. Physiol Rev. 2019;99(3):1325–80.
5. Rechtschaffen A, Bergmann BM. Sleep deprivation and host defense. Am J Physiol Regul Integr Comp Physiol. 2001;280:R602–3.
6. Medzhitov R. Inflammation 2010: new adventures of an old flame. Cell. 2010;140:771–6.
7. Pecchi E, Dallaporta M, Jean A, Thirion S, Troadec JD. Prostaglandins and sickness behavior: old story, new insights. Physiol Behav. 2009;97:279–92.
8. Rosenfeld Y, Shai Y. Lipopolysaccharide (Endotoxin)-host defense antibacterial peptides interactions: role in bacterial resistance and prevention of sepsis. Biochim Biophys Acta. 2006;1758:1513–22.
9. Irwin MR, Wang M, Campomayor CO, Collado Hidalgo A, Cole S. Sleep deprivation and activation of morning levels of cellular and genomic markers of inflammation. Arch Intern Med. 2006;166:1756–62.
10. Brasier AR. The NF-kappaB regulatory network. Cardiovasc Toxicol. 2006;6:111–30.
11. Perkins ND. Integrating cell-signalling pathways with NF-kappaB and IKK function. Nat Rev Mol Cell Biol. 2007;8:49–62.
12. Krueger JM, Obal FJ, Fang J, Kubota T, Taishi P. The role of cytokines in physiological sleep regulation. Ann N Y Acad Sci. 2001;933:211–21.
13. Basheer R, Rainnie DG, Porkka Heiskanen T, Ramesh V, McCarley RW. Adenosine, prolonged wakefulness, and A1-activated NF-kappaB DNA binding in the basal forebrain of the rat. Neuroscience. 2001;104:731–9.
14. Brandt JA, Churchill L, Rehman A, Ellis G, Memet S, Israel A, Krueger JM. Sleep deprivation increases the activation of nuclear factor kappa B in lateral hypothalamic cells. Brain Res. 2004;1004:91–7.
15. Chen Z, Gardi J, Kushikata T, Fang J, Krueger JM. Nuclear factor-kappaB-like activity increases in murine cerebral cortex after sleep deprivation. Am J Phys. 1999;276:R1812–8.
16. Ramesh V, Thatte HS, McCarley RW, Basheer R. Adenosine and sleep deprivation promote NF-kappaB nuclear translocation in cholinergic basal forebrain. J Neurochem. 2007;100:1351–63.
17. Kim J, Hakim F, Kheirandish-Gozal L, et al. Inflammatory pathways in children with insufficient or disordered sleep. Respir Physiol Neurobiol. 2011;178(3):465–74.
18. Imeri L, Opp MR. How (and why) the immune system makes us sleep. Nat Rev Neurosci. 2009;10:199–210.
19. Vgontzas AN, Bixler EO, Chrousos GP. Obesity-related sleepiness and fatigue: the role of the stress system and cytokines. Ann N Y Acad Sci. 2006;1083:329–44.
20. Allan SM, Rothwell NJ. Cytokines and acute neurodegeneration. Nat Rev Neurosci. 2001;2:734–44.
21. Eriksson C, Nobel S, Winblad B, Schultzberg M. Expression of interleukin 1 alpha and beta, and interleukin 1 receptor antagonist mRNA in the rat central nervous system after peripheral administration of lipopolysaccharides. Cytokine. 2000;12:423–31.
22. Krueger JM. The role of cytokines in sleep regulation. Curr Pharm Des. 2008;14:3408–16.
23. Takahashi S, Kapas L, Fang J, Krueger JM. Somnogenic relationships between tumor necrosis factor and interleukin-1. Am J Phys. 1999;276:R1132–40.
24. Spath Schwalbe E, Hansen K, Schmidt F, Schrezenmeier H, Marshall L, Burger K, Fehm HL, Born J. Acute effects of recombinant human interleukin-6 on endocrine and central nervous sleep functions in healthy men. J Clin Endocrinol Metab. 1998;83:1573–9.
25. Ranjbaran Z, Keefer L, Stepanski E, Farhadi A, Keshavarzian A. The relevance of sleep abnormalities to chronic inflammatory conditions. Inflamm Res. 2007;56:51–7.
26. Veler H. Sleep and inflammation bidirectional relationship. Sleep Med Clin. 2023;18:213–8.
27. Lange T, Dimitrov S, Born J. Effects of sleep and circadian rhythm on the human immune system. Ann NY Acad Sci. 2010;1193:48–59.

28. Vgontzas AN, Papanicolaou DA, Hermida RC, et al. Circadian interleukin-6 secretion and quantity and depth of sleep. J Clin Endocrinol Metab. 1999;84:2603–7.
29. Fernandez-Mendoza J, Baker JH, Alexandros NV, et al. Insomnia symptoms with objective short sleep duration are associated with systemic inflammation in adolescents. Brain Behav Immun. 2017;61:110–6.
30. Hasin DS, Sarvet AL, Meyers JL, et al. Epidemiology of adult DSM-5 major depressive disorder and its specifiers in the United States. JAMA Psychiatry. 2018;75(4):336–46.
31. Germain A, Kupfer DJ. Circadian rhythm disturbance in depression. Hum Psychopharmacol. 2008;23(7):571–85.
32. Germain A, Nofzinger EA, Meltzer CC, Wood AM, Kupfer DJ, Moore RY, Buysse DJ. Diurnal variation in regional brain glucose metabolism in depression. Biol Psychiatry. 2007;62:438–45.
33. Buysse DJ, Nofzinger EA, Germain A, Meltzer CC, Wood A, Ombao H, Kupfer DJ, Moore RY. Regional brain glucose metabolism during morning and evening wakefulness in humans: preliminary findings. Sleep. 2004;27:1245–54.
34. Koenigsberg HW, Teicher MH, Mitropoulou V, Navalta C, New AS, Trestman R, Siever LJ. 24-h Monitoring of plasma norepinephrine, MHPG, cortisol, growth hormone and prolactin in depression. J Psychiatr Res. 2004;38:503–11.
35. Almeida OP, Pfaff JJ. Sleep complaints among older general practice patients: association with depression. Br J Gen Pract. 2005;55:864–6.
36. Tsuno N, Besset A, Ritchie K. Sleep and depression. J Clin Psychiatry. 2005;66:1254–69.
37. Saeed SA, Bruce TJ. Seasonal affective disorders. Am Fam Physician. 1998;57:1340–6, 1351–2.
38. Lewy AJ. The dim light melatonin onset, melatonin assays and biological rhythm research in humans. Biol Signals Recept. 1999;8:79–83.
39. Lewy AJ, Bauer VK, Cutler NL, Sack RL, Ahmed S, Thomas KH, Blood ML, Jackson JM. Morning vs evening light treatment of patients with winter depression. Arch Gen Psychiatry. 1998;55:890–6.
40. Lewy AJ, Emens J, Jackman A, Yuhas K. Circadian uses of melatonin in humans. Chronobiol Int. 2006;23:403–12.
41. Lam RW, Levitan RD. Pathophysiology of seasonal affective disorder: a review. J Psychiatry Neurosci. 2000;25:469–80.
42. Sadock BJ, Sadock VA, Pedro R. Kaplan & Sadock's comprehensive textbook of psychiatry. 10th ed. Philadelphia: Wolters Kluwer; 2017.
43. Deltito JA, Moline M, Pollak C, Martin LY, Maremmani I. Effects of phototherapy on non-seasonal unipolar and bipolar depressive spectrum disorder. J Affect Disord. 1991;23(4):231–7.
44. Maes M, Carvalho AF. The Compensatory Immune-Regulatory Reflex System (CIRS) in depression and bipolar disorder. Mol Neurobiol. 2018;55(12):8885–903.
45. Maes M, Lambrechts J, Bosmans E, Jacobs J, Suy E, Vandervorst C, et al. Evidence for a systemic immune activation during depression: results of leukocyte enumeration by flow cytometry in conjunction with monoclonal antibody staining. Psychol Med. 1992;22(1):45–53.
46. Maes M, Bosmans E, Suy E, Vandervorst C, De Jonckheere C, Raus J. Immune disturbances during major depression: upregulated expression of interleukin-2 receptors. Neuropsychobiology. 1990;24(3):115–20.
47. Maes M, Stevens WJ, Declerck LS, Bridts CH, Peters D, Schotte C, et al. Significantly increased expression of T-cell activation markers (interleukin-2 and HLA-DR) in depression: Further evidence for an inflammatory process during that illness. Prog Neuro-Psychopharmacol Biol Psychiatry. 1993;17(2):241–55.
48. Maes M, Scharpé S, Meltzer HY, Bosmans E, Suy E, Calabrese J, et al. Relationships between interleukin-6 activity, acute phase proteins, and function of the hypothalamic-pituitary-adrenal axis in severe depression. Psychiatry Res. 1993;49(1):11–27.
49. Maes M, Bosmans E, Suy E, Vandervorst C, DeJonckheere C, Raus J. Depression-related disturbances in mitogen-induced lymphocyte responses and interleukin-1b and soluble interleukin-2 receptor production. Acta Psychiatr Scand. 1991;84(4):379–86.

50. Maes M, Smith R, Scharpe S. The monocyte-T-lymphocyte hypothesis of major depression. Amsterdam: Elsevier Science; 1995. p. 111–6.
51. Mikova O, Yakimova R, Bosmans E, Kenis G, Maes M. Increased serum tumor necrosis factor alpha concentrations in major depression and multiple sclerosis. Eur Neuropsychopharmacol. 2001;11(3):203–8.
52. Maes M. Acute phase protein alterations in major depression: a review. Rev Neurosci. 1993;4:407–16.
53. Almulla AF, Abbas Abo Algon A, Tunvirachaisakul C, Al-Hakeim HK. Maes M.T helper-1 activation via interleukin-16 is a key phenomenon in the acute phase of severe, first-episode major depressive disorder and suicidal behaviors. J Adv Res. 2024;64:171–81.
54. Çakici N, Sutterland AL, Penninx BWJH, Dalm VA, de Haan L, van Beveren NJM. Altered peripheral blood compounds in drug-naïve first-episode patients with either schizophrenia or major depressive disorder: a meta-analysis. Brain Behav Immun. 2020;88:547–58.
55. Yang K-C, Liu M-N, Liou Y-J, Hu L-Y, Yang B-H, Chou Y-H. Interleukin-1 family and serotonin transporter in first-episode, drug-naive major depressive disorder: A pilot study. J Psychiatr Res. 2021;135:174–80.
56. Maes M, Moraes JB, Bonifacio KL, Barbosa DS, Vargas HO, Michelin AP, et al. Towards a new model and classification of mood disorders based on risk resilience, neuro-affective toxicity, staging, and phenome features using the nomothetic network psychiatry approach. Metab Brain Dis. 2021;36(3):509–21.
57. Maes M, Rachayon M, Jirakran K, Sodsai P, Klinchanhom S, Gałecki P, et al. The immune profile of major dysmood disorder: proof of concept and mechanism using the precision nomothetic psychiatry approach. Cells. 2022;11(7)
58. McEwen BS, Karatsoreos IN. Sleep deprivation and circadian disruption stress, allostasis, and allostatic load. Sleep Med Clin. 2022;17:253–62.
59. Dhaenen H, Boer JA, Willner P. Biological psychiatry. England: Wiley; 2002.
60. Alexopoulos GS, Morimoto SS. The inflammation hypothesis in geriatric depression. Int J Geriatr Psychiatry. 2011;11:1109–18.
61. Lowden A, Moreno C, Holmback U, et al. Eating and shift work-effects on habits, metabolism and performance. Scand J Work Environ Health. 2010;36:150–62.
62. Karatsoreos IN, Bhagat S, Bloss EB, et al. Disruption of circadian clocks has ramifications for metabolism, brain, and behavior. Proc Natl Acad Sci USA. 2011;108:1657–62.
63. Sulli G, et al. Pharmacological activation of REV-ERBs is lethal in cancer and oncogene-induced senescence. Nature. 2018;553:351–5.
64. Ray S, et al. Phenotypic proteomic profiling identifies a landscape of targets for circadian clock-modulating compounds. Life Sci Alliance. 2019;2(6):e201900603.
65. Nomura K, et al. Selective serotonin reuptake inhibitors and raft inhibitors shorten the period of Period1-driven circadian bioluminescence rhythms in rat-1 fibroblasts. Life Sci. 2008;82:1169–74.
66. Sawai Y, et al. In vivo evaluation of the effect of lithium on peripheral circadian clocks by real-time monitoring of clock gene expression in near-freely moving mice. Sci Rep. 2019;9:10909.
67. Sanghani HR, et al. Patient fibroblast circadian rhythms predict lithium sensitivity in bipolar disorder. Mol Psychiatry. 2021;26:5252–65.
68. Palagini L, et al. International expert opinions and recommendations on the use of melatonin in the treatment of insomnia and circadian sleep disturbances in adult neuropsychiatric disorders. Front Psychiatry. 2021;12:688890.
69. Santos P, et al. Anxiolytic properties of compounds that counteract oxidative stress, neuroinflammation and glutamatergic dysfunction: a review. Braz J Psychiatry. 2019;41:168–78.
70. Satyanarayan SK, et al. Melatonergic agonist regulates circadian clock genes and peripheral inflammatory and neuroplasticity markers in patients with depression and anxiety. Brain Behav Immun. 2020;85:142–51.

Inflammatory Impact on Sleep, Schizophrenia, and Autism Spectrum Disorder

Satoshi Kamijo and Hideki Miwa

Abstract This chapter provides an overview of sleep disturbances, psychiatric disorders, including schizophrenia and autism spectrum disorder (ASD), and how inflammation may be involved. We will explore scientific evidence whether sleep disturbances and inflammation contribute to an increased risk of developing psychiatric disorders. In conclusion, we find the relationship between psychiatric disorders and sleep is complex and includes two-way causality, although the underlying mechanisms are still not fully understood.

Keywords Maternal immune activation · Autoimmune disease · Maternal infection · Pro-inflammatory cytokines · Microglial activation · Brain-reactive autoantibodies · Sleep disturbances · Schizophrenia · Autism spectrum disorder

1 Introduction

The immune system and the nervous system are closely linked, with two-way interactions between them. In the brain's immune system, two types of glial cells—microglia and astrocytes—play key roles. It has been repeatedly reported that infection during pregnancy or early postnatal development increases the risk of developing psychiatric diseases like schizophrenia and autism spectrum disorder (ASD), in later life [1]. A large-scale study from Denmark also found that a history of infections in mothers and infants is linked to a higher risk of psychiatric disorder [2–5]. Additionally, abnormal sleep patterns—such as being a "night owl", shorter sleep duration, and insomnia—are more common in patients with these psychiatric disorders. Sleep disturbances are also linked to higher risk of infectious diseases,

S. Kamijo · H. Miwa (✉)
Molecular Neuropsychopharmacology Section, Department of Neuropsychopharmacology, National Institute of Mental Health, National Center of Neurology and Psychiatry, Tokyo, Japan
e-mail: Hideki_Miwa@ncnp.go.jp

P. K. Jha, H. Bouaouda (eds.), *Circadian Rhythms, Sleep and Inflammation*, Progress in Inflammation Research 93,
https://doi.org/10.1007/978-3-031-74744-1_7

suggesting that sleep might have a role in the regulation of anti-viral immune responses. Interestingly, sleep disturbances are associated with an increased risk of developing psychiatric disorders, creating a complex, two-way relationship between psychiatric disorders and sleep. However, the exact mechanisms behind this link are still unknown. In this chapter, we provide an overview of the current understanding of sleep disturbances, psychiatric disorders, including schizophrenia and ASD, and the potential role of inflammation.

2 Reciprocal Interaction Between Sleep and Inflammation

Many of you have probably experienced the feeling of fatigue and drowsiness when you catch a cold. This serves as a clear example of how acute inflammation can impact our sleep patterns. These physical responses against infection, referred to as "sickness behavior," are physiologically reasonable [6]; They help conserve energy, allowing the body to focus on fighting the infection and promoting recovery. However, the effects of acute inflammation on human sleep behavior are not consistent across studies, mainly due to the fact that the subjective impact of symptoms caused by the infection, which can influence sleep quality differently [7]. Additionally, there are few studies examining sleep behavior during real infections in humans because of ethical limitaions.

In studies where influenza A/B and the common cold (rhinovirus-type-23) were experimentally induced, participants reported longer sleep duration not only when they had the symptoms but also during the incubation phase [8]. However, in a different study where healthy people were exposed to rhinovirus-type-23, objective measurements using polysomnography showed reduced total sleep duration and sleep efficiency when symptoms were active [9]. Addictional insights into how inflammation affects sleep come from studies involving subjects vaccinated or injected with lipopolysaccharides (LPS), which activates the immune systems without introducing actual pathogens. For example, after receiving a typhoid vaccine, subjects had more nighttime awakenings as measured by polysomnography on the night following vaccination [10]. LPS endotoxin, a major component of the cell walls of Gram-negative bacteria, is known to trigger immune activation. People injected with LPS showed longer duration and more intense periods of non-rapid eye movement (NREM) sleep when their inflammatory cytokine elevation and physical symptoms were moderate; however, as inflammation became more intense, the duration and intensity of NREM sleep decreased [11–13].

Rheumatoid arthritis (RA) is a typical example of chronic inflammatory disease. Sleep disturbances are frequently reported in RA patients, but research findings on this issue have been inconsistent. Only a few studies have investigated how RA specifically affects sleep. Infliximab, a potent antagonizing antibody against pro-inflammatory cytokine TNF-α, is used to treat RA. At least three studies using polysomnography found that RA patients undergoing infliximab therapy experienced improvements in sleep duration and efficiency[14–16]. However, in a large cohort study, no significant difference was observed in sleep patterns between those receiving anti-TNF-α therapy and those who did not [17]. Another drug tocilizumab, an antibody that blocks the receptor for pro-inflammatory cytokine IL-6, has also been linked to self-reported improvements in sleep quality among RA patients [18]. These findings suggest that pro-inflammatory cytokines are responsible for sleep disturbances among RA patients. Interestingly, some studies indicate that sleep improvements are not directly related to the alleviation of the physical symptoms and the observed effect is not attributed to pain relief [14]. However, many studies failed to adequately account for confounding factors like disease activity, pain status, and comorbid disorders. These limitations make it difficult about how RA treatments affect sleep.

The physiological influence of normal sleep on the immune system has been explored through sleep deprivation experiments. A meta-analysis of human studies revealed that acute sleep deprivation or chronic sleep restriction did not increase levels of peripheral inflammatory markers in healthy individuals [19]. However, while blood cytokine level remained unchanged, peripheral leukocytes showed the increased expression of upstream pro-inflammatory transcription factors [20]. Under chronic sleep disturbance conditions, a meta-analysis of the observational studies found an association between sleep disruption and increased levels of pro-inflammatory cytokine IL-6 and inflammatory response indicator C-reactive protein (CRP) [19]. Furthermore, a longitudinal prospective study demonstrated that sleep disruption reliably predicted elevated IL-6 and CRP levels [21]. These apparently conflicting results between acute and chronic sleep disturbances can be partially explained by the difference in the duration of sleep disruption. The relationship between elevated inflammatory cytokines and chronic sleep disturbance should be interpreted with caution, as the underlying inflammatory disease itself could be the primary cause of sleep disorder. However, a randomized controlled trial involving patients with insomnia demonstrated that treating insomnia led to a decrease in peripheral inflammatory cytokine levels, suggesting that chronic sleep disturbance can contribute to a pro-inflammatory state [22].

In summary, peripheral inflammation can influence sleep patterns in various ways, likely through the action of inflammatory cytokines. On the other hand, normal sleep helps regulate inflammation as evidenced by the fact that chronic sleep disturbances exacerbate inflammation. Recent reports have linked chronic inflammation not only to diseases like heart disease and diabetes but also to conditions such as aging and obesity [23]. Understanding how chronic inflammation and sleep disturbances interact is crucial. Exploring how these two factors influence each other and contribute together to the development of chronic diseases and conditions could provide valuable insights for prevention and treatment.

3 Impact of Inflammation on Psychiatric Diseases

It has been repeatedly reported that infection during pregnancy or early postnatal development can increases the risk of developing psychiatric diseases later in life [1]. For example, children born during influenza endemics have been found to have a higher likelihood of being diagnosed with schizophrenia in adulthood [24, 25]. Additionally, findings from a large-scale Danish population study suggest that a history of maternal or postnatal infections is associated with a higher risk of developing psychiatric diseases in offspring [2–5].

3.1 Influence of Maternal Inflammation on Psychiatric Diseases

In most cases, maternal inflammation results from infection. Several mechanisms have been proposed to explain how maternal inflammation impacts fetal development [26]. First, infections can damage the placenta, causing fetal distress. Second, cytokines and antibodies produced due to maternal immune activation can traverse the placental barrier and then affect the developing fetus. Third, maternally-produced cytokines and other humoral factors can prime the fetal microglial cells into a ready-to-go state. Lastly, these transplacental effects may lead to epigenetic modifications in the fetal genome. The link between maternal infection and an increased susceptibility to psychiatric disorders has been well-documented. However, the exact critical period remains a topic of ongoing debate. A large Danish cohort study, which included over 1.2 million individuals, unveiled a significant correlation between the treatment history of infections during pregnancy and an elevated risk of developing psychiatric diseases in offspring; Specifically, the relative increase in risk for all psychiatric diseases was 1.21. Even in cases where only medications were used to treat milder symptoms, the relative risk increase was 1.09 [4]. Furthermore, infections during the second and third trimesters were associated with a greater prevalence of psychiatric disorders in their offspring. Interestingly, the study found little difference in risk between bacterial and viral pathogens. Importantly, further analysis comparing the periods before, during, and after pregnancy revealed that the elevated risk persisted, suggesting that while maternal infection is a contributing factor, genetic susceptibility to infections may play a significant role in the observed outcomes.

3.2 *Influence of Postnatal Inflammation on Psychiatric Diseases*

Another study, also utilizing the comprehensive Danish nationwide dataset, found that the treatment history of postnatal infections has shown an association with an elevated risk of developing psychiatric disorders later in life [3]. Specifically, infections requiring hospitalization were associated with a 1.84-fold higher likelihood of being diagnosed with a mental disorder, while infections treated only with anti-infective medications raised the risk by 1.40-fold. This investigation focused on mental disorders manifesting in children and adolescents because the maximum follow-up age at the time of publication was 18 years old. Consequently, the findings are less sensitive to psychiatric conditions that typically emerge in adulthood. Despite this limitation, the study underscored the increased risk of psychiatric disorders associated with postnatal infections, with a clear dose-dependent relationship: a higher number of infections correlated with a greater risk. While infections can inflict direct damage on the central nervous system (CNS), as seen in conditions like meningitis or encephalitis, several indirect mechanisms are also possible. First, cytokines produced peripherally during infection can infiltrate the CNS, priming microglial cells. Second, the antibiotics used to treat the infection may alter the balance of gut flora, potentially influencing brain health. Third, as mentioned above, genetic susceptibility to infections may be associated with the risk loci of psychiatric disorders.

4 Schizophrenia and Inflammation

4.1 *Effects of Maternal Inflammation on Schizophrenia*

The link between schizophrenia and inflammation has undergone a comprehensive investigation [27, 28]. It is reported that various infectious diseases during pregnancy elevate the risk of schizophrenia [24]. The most well-studied risk factor is influenza. The studies of the influenza epidemic during the 1950s revealed a noteworthy association—children born during the epidemic years had a significantly higher likelihood of being diagnosed with schizophrenia [29–34], while other studies showed no association [35, 36]. According to a comprehensive review, maternal influenza during pregnancy increased the odds of schizophrenia in offspring by threefold, accounting for approximately 14% of the patient population [24]. However, a subsequent meta-analysis found no heightened risk in offspring exposed to maternal influenza, and this study faced critique for its methodology [37, 38], underscoring the ongoing debate surrounding the extent of risk increase attributed to maternal influenza. The increased risk of schizophrenia is reported to be associated with elevated levels of maternal pro-inflammatory cytokine IL-8 [39]. This result implies that maternal immune activation alone is sufficient for raising the risk, independent of specific etiologies. In animal models, both viral and bacterial agents can induce schizophrenia-like phenotypes, measured using a prepulse inhibition test that reflects abnormal responses akin to those observed in human schizophrenia patients [40]. However, despite this body of evidence, a Danish nationwide study unexpectedly revealed that maternal infection during pregnancy did not elevate the risk of schizophrenia [4]. The population of the study remained relatively young, and the follow-up period might be insufficient to detect conditions like schizophrenia, which typically emerge in adulthood. Additionally, maternal autoimmune diseases are also an independent risk factor for schizophrenia, increasing the risk by 1.35 times [41].

4.2 Effects of Postnatal Inflammation on Schizophrenia

In the realm of adult epidemiological data, a history of autoimmune disease increases the incidence rate ratio of schizophrenia by 1.29 times, while hospitalization due to infections raises the risk by 1.60 times[2]. When both factors coexist, the incidence rate ratio further to 2.25 times, with infection frequency leading to an even greater increase in a dose-dependent manner. Infections of the central nervous system (CNS) during adulthood also elevate the likelihood of being diagnosed with schizophrenia spectrum disorders by 1.23 times. [5]. Notably, hospitalization due to infections during earlier life stages elevate the incident rate ratio by 1.80 times [3]. These findings collectively highlight the association between autoimmune disease history, infections across various life stages, and the increased incidence of schizophrenia. Further investigations into the immune status of schizophrenia patients have revealed significant insights. Postmortem studies have revealed increased inflammatory responses in the brains of schizophrenia patients [42]. Positron emission tomography (PET) imaging of the first-episode and chronic schizophrenia patients highlighted the activation of microglia in their brains [43–45]. Additionally, serum levels of C-reactive protein (CRP) and IL-6 were found to be elevated in schizophrenia patients [46, 47]. Genomic investigations have further identified an association between immune-related genes, such as HLA (human leukocyte antigen) and complement C4 with the disease risk [48–51].

A pivotal question emerges from these findings: Is the inflammation a causative factor for schizophrenia or merely an epiphenomenon? A double-blind randomized controlled study examined the impact of treating schizophrenia patients with the anti-inflammatory agent celecoxib, a potent COX-2 inhibitor; first-episode schizophrenia patients treated with a combination of antipsychotics and celecoxib exhibited significant improvements in psychological symptoms compared to those treated solely with antipsychotics [52, 53], suggesting that inflammation is not an epiphenomenon but rather a critical factor in the pathogenesis of schizophrenia. Nevertheless, subsequent follow-up studies yielded inconsistent results [54]. While chronic schizophrenia patients treated with celecoxib showed no statistically significant effects [55], the other study demonstrated the benefits of celecoxib treatment for acute and chronic schizophrenia symptoms, including both positive and negative symptoms [56]. These discrepancies may be explained by differences in the timing of treatment, with patients in earlier stages of the disease showing greater responsiveness.

In summary, epidemiological evidence links immune activation to an increased incidence of schizophrenia, spanning maternal, childhood, and adulthood infections. Patients with schizophrenia represented elevated immune response and their microglia are activated. These results implied the involvement of the immune system in schizophrenia pathogenesis. Genomic studies revealed that both innate and adaptive immune system-related genes were associated with schizophrenia. Although the results were inconsistent, anti-inflammatory treatment was effective in improving the psychological symptoms of schizophrenia, suggesting that inflammation critically contributes to the development of the disease.

5 ASD and Inflammation

5.1 Effects of Maternal Inflammation on ASD

The relationship between maternal inflammation and ASD has been the subject of extensive investigations [26, 57, 58]. Notably, a comprehensive Danish nationwide study has shed light on a significant association between maternal infections and an increased risk of ASD in their offspring [4]. The history of maternal infections requiring hospitalization during pregnancy was linked to a 1.22-fold rise in the risk of ASD. Other studies have connected maternal infections such as rubella, measles, mumps, cytomegalovirus, polyomavirus, influenza, and bacterial infections during pregnancy to an elevated risk of ASD. However, the specific stage of pregnancy when these infections might have the most significant impact on ASD development remains unclear. Beyond infectious factors, chronic inflammatory conditions like asthma and obesity have also emerged as potential contributors to an increased risk of ASD [59–61]. Addressing whether maternal immune activation alone can induce ASD-like phenotypes, researchers have developed animal models to evaluate the effects of maternal infection on fetal brain development. Specifically, injection of synthetic double-stranded RNA poly I:C into the pregnant subjects to mimic viral infection has been shown to trigger maternal immune responses capable of inducing ASD-like behaviors in their offspring, such as reduced sociability and repetitive behavior [62–65]. This poly I:C model has been validated across diverse species, including mice, rats, marmosets, and macaques, establishing maternal immune activation as a universal ASD risk factor [66, 67]. These ASD-like phenotypes were also observed in the offspring of mothers receiving a single injection of pro-inflammatory cytokine IL-6 during pregnancy. Strikingly, this phenomenon was not observed in the offspring of IL-1α, TNF-α, or interferon-γ injected mothers [68, 69]. Moreover, these abnormal behavioral phenotypes of IL-6-injected offspring were rescued with the co-injection of cytokine-neutralizing antibodies [69]. Additionally, the offspring of IL-6 knockout mice subjected to poly I:C injections displayed no behavioral deviations when compared to non-injected controls [70]. These findings prove that it is not the pathogens themselves but the maternal immune activation that plays a pivotal role in ASD pathogenesis, exerting its influence primarily through inflammatory cytokines.

Maternal autoimmune disease constitutes another independent risk factor for ASD, as indicated by meta-analysis [71] and a nationwide cohort study [41]. Notably, it is reported that ASD children's mothers are four times more likely to possess antibodies against brain tissue [72]. Validating the pathogenic potential of brain-reactive antibodies, mice injected with immunoglobulins collected from ASD children's mothers showed abnormal behavioral phenotypes [73]. These findings suggest that the transplacental transfer of brain-reactive autoantibodies may significantly influence fetal brain development.

5.2 *Effects of Postnatal Inflammation on ASD*

The link between a history of infections and an increased risk of ASD has been widely documented [3, 5]. However, its causal relationship remains elusive. Studies from various countries identified certain HLA genes, especially class II DRB1, as being associated with a higher risk of ASD [74–76]. Genome-wide association studies, similar to those conducted for other psychiatric disorders, have yielded complex results [77–79]. Notably, genes that are expressed differently in autistic brains compared to control brains are often involved in immune-related functions [80]. Dozens of studies have explored the systemic inflammation status in patients with ASD [81–85]. While the results vary, heightened levels of inflammatory cytokines have been consistently observed; Plasma IL-1β and IL-6, which are pro-inflammatory cytokines, were significantly increased in adults with severe ASD [86]. In a cohort of ASD children, elevated levels of pro-inflammatory cytokines have been reported, with their severity correlating with behavioral phenotypes' intensity [87]. In addition, histological evidence underscores local inflammation within the brain of patients with ASD. Immunohistological staining of HLA-DR, a marker of microglial cells, reveals increased expression levels in the cerebral cortex, white matter, and most prominently cerebellum, in the postmortem brains of patients with ASD [88]. Moreover, the number of activated microglia has been observed to be increased in the dorsolateral prefrontal cortex in patients with ASD [89]. This phenomenon has also been observed in an ASD model exposed to valproic acid, indicating that the potential universality of local CNS inflammation as a shared pathological mechanism of ASD [90]. Beyond immune response alterations, dysregulation of the gut flora in patients with ASD has been repeatedly pointed out [91–93]. In ASD mouse models, germ-free mice exhibit reduced sociability, and their phenotype can be recovered through transplantation of the gut flora of typically developing mice [94]. This finding suggests that gut flora can modulate the immune system and impact normal brain development, albeit the mechanisms underlying these effects are only beginning to be understood. While studies assessing autoimmune disease in children with ASD are limited due to the delayed onset of autoimmune disease in most cases, it's noteworthy that children with ASD are more likely to possess brain-reactive autoantibodies compared to typically developing control subjects [95, 96].

In summary, maternal inflammation emerges as a contributor to elevated ASD risk in offspring. Maternal immune activation alone could induce ASD phenotypes and its influence is likely to be mediated by inflammatory cytokines. Patients with ASD predominantly exhibit a pro-inflammatory immune profile, often coupled with dysregulated gut flora. Brain-reactive autoantibodies are frequently identified in both patients with ASD and their mothers. Genomic investigations have highlighted numerous immune-related genes as ASD risk factors and histological analysis of the postmortem brains revealed microglial activation in ASD patients. The presence of local microglial activation may well constitute a shared mechanism underpinning the pathogenesis of ASD.

6 Conclusion

We briefly discussed the current understanding of how inflammation affects sleep and psychiatric disease. Inflammation can disrupts normal sleep behaviors and in turn, disrupted sleep can activate inflammatory responses. Human epidemiological data have revealed that both maternal and postnatal inflammation are associated with an increased risk of developing psychiatric disorders later in life. Prenatal and postnatal inflammatory risk factors of psychiatric diseases and their mode of action are summarized in Fig. 1. Patients with schizophrenia and ASD exhibit elevated levels of peripheral pro-inflammatory cytokines and activated microglia. These findings suggest that inflammation may play a causal role in the development of these disorders. Supporting this, animal studies have demonstrated that immune activation alone can lead to abnormal behavioral traits.

It is widely known that patients with psychiatric disorders often experience sleep disturbances more frequently. Inflammation could serve as a common underlying factor linking sleep disturbances with psychiatric disorders. Dysregulated sleep and inflammation mutually influence each other, creating a "vicious circle" that modulates the progression of psychiatric disease.

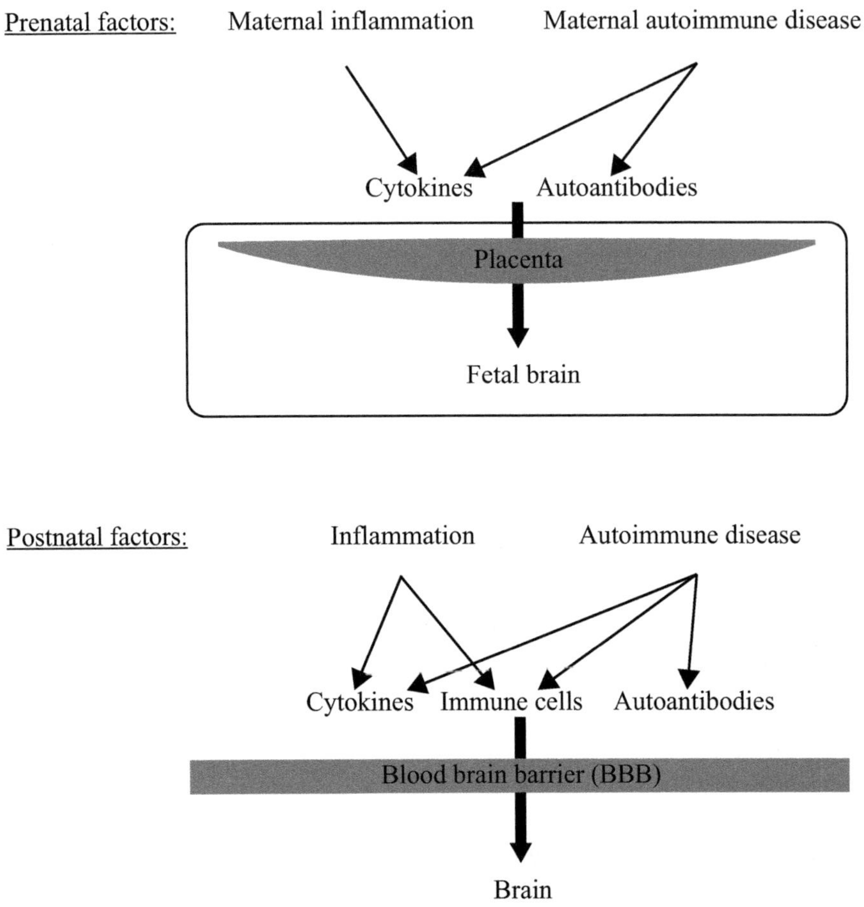

Fig. 1 Prenatal and postnatal inflammatory risk factors of psychiatric diseases. Maternal inflammation alters fetal brain development by transplacental transfer of pro-inflammatory cytokines. Maternal autoantibodies can also cross the placental barrier to disturb normal development. Postnatal inflammatory diseases are independent risk factors of psychiatric diseases. Peripheral inflammatory status is transmitted to CNS via cytokines, autoantibodies, translocation of immune cells, and local activation of microglia

Compliance with Ethical Standards The authors declare that they have no conflict of interest regarding the publication of this book chapter. The authors would like to thank the National Center of Neurology and Psychiatry, Japan for their support through Intramural Research Grant (6-1) for Neurological and Psychiatric Disorders (to HM) during the preparation of this review.

References

1. Pape K, Tamouza R, Leboyer M, Zipp F. Immunoneuropsychiatry—novel perspectives on brain disorders. Nat Rev Neurol. 2019;15(6):317–28.
2. Benros ME, Nielsen PR, Nordentoft M, Eaton WW, Dalton SO, Mortensen PB. Autoimmune diseases and severe infections as risk factors for schizophrenia: a 30-year population-based register study. Am J Psychiatry. 2011;168(12):1303–10.
3. Köhler-Forsberg O, Petersen L, Gasse C, Mortensen PB, Dalsgaard S, Yolken RH, et al. A nationwide study in denmark of the association between treated infections and the subsequent risk of treated mental disorders in children and adolescents. JAMA Psychiatry. 2019;76(3):271–9.
4. Lydholm CN, Köhler-Forsberg O, Nordentoft M, Yolken RH, Mortensen PB, Petersen L, et al. Parental infections before, during, and after pregnancy as risk factors for mental disorders in childhood and adolescence: a nationwide Danish study. Biol Psychiatry. 2019;85(4):317–25.
5. Pedersen EMJ, Köhler-Forsberg O, Nordentoft M, Christensen RHB, Mortensen PB, Petersen L, et al. Infections of the central nervous system as a risk factor for mental disorders and cognitive impairment: A nationwide register-based study. Brain Behav Immun. 2020;88:668–74.
6. Dantzer R, Kelley KW. Twenty years of research on cytokine-induced sickness behavior. Brain Behav Immun. 2007;21(2):153–60.
7. Irwin MR. Sleep and inflammation: partners in sickness and in health. Nat Rev Immunol. 2019;19(11):702–15.
8. Besedovsky L, Lange T, Haack M. The sleep-immune crosstalk in health and disease. Physiol Rev. 2019;99(3):1325–80.
9. Drake CL, Roehrs TA, Royer H, Koshorek G, Turner RB, Roth T. Effects of an experimentally induced rhinovirus cold on sleep, performance, and daytime alertness. Physiol Behav. 2000;71(1–2):75–81.
10. Sharpley AL, Cooper CM, Williams C, Godlewska BR, Cowen PJ. Effects of typhoid vaccine on inflammation and sleep in healthy participants: a double-blind, placebo-controlled, crossover study. Psychopharmacology (Berl). 2016;233(18):3429–35.
11. Pollmächer T, Schreiber W, Gudewill S, Vedder H, Fassbender K, Wiedemann K, et al. Influence of endotoxin on nocturnal sleep in humans. Am J Physiol. 1993;264(6 Pt 2):R1077–83.
12. Mullington J, Korth C, Hermann DM, Orth A, Galanos C, Holsboer F, et al. Dose-dependent effects of endotoxin on human sleep. Am J Physiol Regul Integr Comp Physiol. 2000;278(4):R947–55.
13. Haack M, Schuld A, Kraus T, Pollmächer T. Effects of sleep on endotoxin-induced host responses in healthy men. Psychosom Med. 2001;63(4):568–78.
14. Zamarrón C, Maceiras F, Mera A, Gómez-Reino JJ. Effect of the first infliximab infusion on sleep and alertness in patients with active rheumatoid arthritis. Ann Rheum Dis. 2004;63(1):88–90.
15. Taylor-Gjevre RM, Gjevre JA, Nair BV, Skomro RP, Lim HJ. Improved sleep efficiency after anti-tumor necrosis factor α therapy in rheumatoid arthritis patients. Ther Adv Musculoskelet Dis. 2011;3(5):227–33.
16. Detert J, Dziurla R, Hoff P, Gaber T, Klaus P, Bastian H, et al. Effects of treatment with etanercept versus methotrexate on sleep quality, fatigue and selected immune parameters in patients with active rheumatoid arthritis. Clin Exp Rheumatol. 2016;34(5):848–56.
17. Wolfe F, Michaud K, Li T. Sleep disturbance in patients with rheumatoid arthritis: evaluation by medical outcomes study and visual analog sleep scales. J Rheumatol. 2006;33(10):1942–51.
18. Fragiadaki K, Tektonidou MG, Konsta M, Chrousos GP, Sfikakis PP. Sleep disturbances and interleukin 6 receptor inhibition in rheumatoid arthritis. J Rheumatol. 2012;39(1):60–2.
19. Irwin MR, Olmstead R, Carroll JE. Sleep disturbance, sleep duration, and inflammation: a systematic review and meta-analysis of cohort studies and experimental sleep deprivation. Biol Psychiatry. 2016;80(1):40–52.

20. Irwin MR, Wang M, Campomayor CO, Collado-Hidalgo A, Cole S. Sleep deprivation and activation of morning levels of cellular and genomic markers of inflammation. Arch Intern Med. 2006;166(16):1756–62.
21. Cho HJ, Seeman TE, Kiefe CI, Lauderdale DS, Irwin MR. Sleep disturbance and longitudinal risk of inflammation: Moderating influences of social integration and social isolation in the Coronary Artery Risk Development in Young Adults (CARDIA) study. Brain Behav Immun. 2015;46:319–26.
22. Irwin MR, Olmstead R, Breen EC, Witarama T, Carrillo C, Sadeghi N, et al. Cognitive behavioral therapy and tai chi reverse cellular and genomic markers of inflammation in late-life insomnia: a randomized controlled trial. Biol Psychiatry. 2015;78(10):721–9.
23. Furman D, Campisi J, Verdin E, Carrera-Bastos P, Targ S, Franceschi C, et al. Chronic inflammation in the etiology of disease across the life span. Nat Med. 2019;25(12):1822–32.
24. Brown AS, Derkits EJ. Prenatal infection and schizophrenia: a review of epidemiologic and translational studies. Am J Psychiatry. 2010;167(3):261–80.
25. Kępińska AP, Iyegbe CO, Vernon AC, Yolken R, Murray RM, Pollak TA. Schizophrenia and influenza at the centenary of the 1918-1919 Spanish influenza pandemic: mechanisms of psychosis risk. Front Psychiatry. 2020;11:72.
26. Han VX, Patel S, Jones HF, Dale RC. Maternal immune activation and neuroinflammation in human neurodevelopmental disorders. Nat Rev Neurol. 2021;17(9):564–79.
27. Khandaker GM, Cousins L, Deakin J, Lennox BR, Yolken R, Jones PB. Inflammation and immunity in schizophrenia: implications for pathophysiology and treatment. Lancet Psychiatry. 2015;2(3):258–70.
28. Müller N, Weidinger E, Leitner B, Schwarz MJ. The role of inflammation in schizophrenia. Front Neurosci. 2015;9:372.
29. McGrath J, Castle D. Does influenza cause schizophrenia? A five year review. Aust N Z J Psychiatry. 1995;29(1):23–31.
30. Mednick SA, Machon RA, Huttunen MO, Bonett D. Adult schizophrenia following prenatal exposure to an influenza epidemic. Arch Gen Psychiatry. 1988;45(2):189–92.
31. Mednick SA, Huttunen MO, Machón RA. Prenatal influenza infections and adult schizophrenia. Schizophr Bull. 1994;20(2):263–7.
32. O'Callaghan E, Sham P, Takei N, Glover G, Murray RM. Schizophrenia after prenatal exposure to 1957 A2 influenza epidemic. Lancet. 1991;337(8752):1248–50.
33. Adams W, Kendell RE, Hare EH, Munk-Jørgensen P. Epidemiological evidence that maternal influenza contributes to the aetiology of schizophrenia. An analysis of Scottish, English, and Danish data. Br J Psychiatry. 1993;163:522–34.
34. Kunugi H, Nanko S, Takei N, Saito K, Hayashi N, Kazamatsuri H. Schizophrenia following in utero exposure to the 1957 influenza epidemics in Japan. Am J Psychiatry. 1995;152(3):450–2.
35. Erlenmeyer-Kimling L, Folnegović Z, Hrabak-Zerjavić V, Borcić B, Folnegović-Smalc V, Susser E. Schizophrenia and prenatal exposure to the 1957 A2 influenza epidemic in Croatia. Am J Psychiatry. 1994;151(10):1496–8.
36. Susser E, Lin SP, Brown AS, Lumey LH, Erlenmeyer-Kimling L. No relation between risk of schizophrenia and prenatal exposure to influenza in Holland. Am J Psychiatry. 1994;151(6):922–4.
37. Selten JP, Termorshuizen F. The serological evidence for maternal influenza as risk factor for psychosis in offspring is insufficient: critical review and meta-analysis. Schizophr Res. 2017;183:2–9.
38. Brown AS, Meyer U. Comment on: "The serological evidence for maternal influenza as risk factor for psychosis in offspring is insufficient: Critical review and meta-analysis". Schizophr Res. 2017;189:223–4.
39. Brown AS, Hooton J, Schaefer CA, Zhang H, Petkova E, Babulas V, et al. Elevated maternal interleukin-8 levels and risk of schizophrenia in adult offspring. Am J Psychiatry. 2004;161(5):889–95.

40. Swerdlow NR, Weber M, Qu Y, Light GA, Braff DL. Realistic expectations of prepulse inhibition in translational models for schizophrenia research. Psychopharmacology (Berl). 2008;199(3):331–88.
41. He H, Yu Y, Liew Z, Gissler M, László KD, Valdimarsdóttir UA, et al. Association of maternal autoimmune diseases with risk of mental disorders in offspring in Denmark. JAMA Netw Open. 2022;5(4):e227503.
42. Körschenhausen DA, Hampel HJ, Ackenheil M, Penning R, Müller N. Fibrin degradation products in post mortem brain tissue of schizophrenics: a possible marker for underlying inflammatory processes. Schizophr Res. 1996;19(2–3):103–9.
43. van Berckel BN, Bossong MG, Boellaard R, Kloet R, Schuitemaker A, Caspers E, et al. Microglia activation in recent-onset schizophrenia: a quantitative (R)-[11C]PK11195 positron emission tomography study. Biol Psychiatry. 2008;64(9):820–2.
44. Doorduin J, de Vries EF, Willemsen AT, de Groot JC, Dierckx RA, Klein HC. Neuroinflammation in schizophrenia-related psychosis: a PET study. J Nucl Med. 2009;50(11):1801–7.
45. Takano A, Arakawa R, Ito H, Tateno A, Takahashi H, Matsumoto R, et al. Peripheral benzodiazepine receptors in patients with chronic schizophrenia: a PET study with [11C]DAA1106. Int J Neuropsychopharmacol. 2010;13(7):943–50.
46. Goldsmith DR, Rapaport MH, Miller BJ. A meta-analysis of blood cytokine network alterations in psychiatric patients: comparisons between schizophrenia, bipolar disorder and depression. Mol Psychiatry. 2016;21(12):1696–709.
47. Miller BJ, Culpepper N, Rapaport MH. C-reactive protein levels in schizophrenia: a review and meta-analysis. Clin Schizophr Relat Psychoses. 2014;7(4):223–30.
48. Trowsdale J, Knight JC. Major histocompatibility complex genomics and human disease. Annu Rev Genomics Hum Genet. 2013;14:301–23.
49. Sekar A, Bialas AR, de Rivera H, Davis A, Hammond TR, Kamitaki N, et al. Schizophrenia risk from complex variation of complement component 4. Nature. 2016;530(7589):177–83.
50. Dendrou CA, Petersen J, Rossjohn J, Fugger L. HLA variation and disease. Nat Rev Immunol. 2018;18(5):325–39.
51. Prasad KM, Chowdari KV, D'Aiuto LA, Iyengar S, Stanley JA, Nimgaonkar VL. Neuropil contraction in relation to Complement C4 gene copy numbers in independent cohorts of adolescent-onset and young adult-onset schizophrenia patients-a pilot study. Transl Psychiatry. 2018;8(1):134.
52. Müller N, Riedel M, Scheppach C, Brandstätter B, Sokullu S, Krampe K, et al. Beneficial antipsychotic effects of celecoxib add-on therapy compared to risperidone alone in schizophrenia. Am J Psychiatry. 2002;159(6):1029–34.
53. Müller N, Krause D, Dehning S, Musil R, Schennach-Wolff R, Obermeier M, et al. Celecoxib treatment in an early stage of schizophrenia: results of a randomized, double-blind, placebo-controlled trial of celecoxib augmentation of amisulpride treatment. Schizophr Res. 2010;121(1–3):118–24.
54. Sethi R, Gómez-Coronado N, Walker AJ, Robertson OD, Agustini B, Berk M, et al. Neurobiology and therapeutic potential of cyclooxygenase-2 (COX-2) inhibitors for inflammation in neuropsychiatric disorders. Front Psychiatry. 2019;10:605.
55. Rapaport MH, Delrahim KK, Bresee CJ, Maddux RE, Ahmadpour O, Dolnak D. Celecoxib augmentation of continuously ill patients with schizophrenia. Biol Psychiatry. 2005;57(12):1594–6.
56. Akhondzadeh S, Tabatabaee M, Amini H, Ahmadi Abhari SA, Abbasi SH, Behnam B. Celecoxib as adjunctive therapy in schizophrenia: a double-blind, randomized and placebo-controlled trial. Schizophr Res. 2007;90(1–3):179–85.
57. McLellan J, Kim DHJ, Bruce M, Ramirez-Celis A, Van de Water J. Maternal immune dysregulation and autism-understanding the role of cytokines, chemokines and autoantibodies. Front Psychiatry. 2022;13:834910.
58. Majerczyk D, Ayad EG, Brewton KL, Saing P, Hart PC. Systemic maternal inflammation promotes ASD via IL-6 and IFN-γ. Biosci Rep. 2022;42(11)

59. Hisle-Gorman E, Susi A, Stokes T, Gorman G, Erdie-Lalena C, Nylund CM. Prenatal, perinatal, and neonatal risk factors of autism spectrum disorder. Pediatr Res. 2018;84(2):190–8.
60. Gong T, Lundholm C, Rejnö G, Bölte S, Larsson H, D'Onofrio BM, et al. Parental asthma and risk of autism spectrum disorder in offspring: A population and family-based case-control study. Clin Exp Allergy. 2019;49(6):883–91.
61. Lei XY, Li YJ, Ou JJ, Li YM. Association between parental body mass index and autism spectrum disorder: a systematic review and meta-analysis. Eur Child Adolesc Psychiatry. 2019;28(7):933–47.
62. Urakubo A, Jarskog LF, Lieberman JA, Gilmore JH. Prenatal exposure to maternal infection alters cytokine expression in the placenta, amniotic fluid, and fetal brain. Schizophr Res. 2001;47(1):27–36.
63. Gilmore JH, Jarskog LF, Vadlamudi S. Maternal poly I:C exposure during pregnancy regulates TNF alpha, BDNF, and NGF expression in neonatal brain and the maternal-fetal unit of the rat. J Neuroimmunol. 2005;159(1–2):106–12.
64. Meyer U, Feldon J, Schedlowski M, Yee BK. Towards an immuno-precipitated neurodevelopmental animal model of schizophrenia. Neurosci Biobehav Rev. 2005;29(6):913–47.
65. Meyer U, Feldon J. Neural basis of psychosis-related behaviour in the infection model of schizophrenia. Behav Brain Res. 2009;204(2):322–34.
66. Bucknor MC, Gururajan A, Dale RC, Hofer MJ. A comprehensive approach to modeling maternal immune activation in rodents. Front Neurosci. 2022;16:1071976.
67. Santana-Coelho D, Layne-Colon D, Valdespino R, Ross CC, Tardif SD, O'Connor JC. Advancing autism research from mice to marmosets: behavioral development of offspring following prenatal maternal immune activation. Front Psychiatry. 2021;12:705554.
68. Meltzer A, Van de Water J. The role of the immune system in autism spectrum disorder. Neuropsychopharmacology. 2017;42(1):284–98.
69. Smith SE, Li J, Garbett K, Mirnics K, Patterson PH. Maternal immune activation alters fetal brain development through interleukin-6. J Neurosci. 2007;27(40):10695–702.
70. Hsiao EY, McBride SW, Hsien S, Sharon G, Hyde ER, McCue T, et al. Microbiota modulate behavioral and physiological abnormalities associated with neurodevelopmental disorders. Cell. 2013;155(7):1451–63.
71. Chen SW, Zhong XS, Jiang LN, Zheng XY, Xiong YQ, Ma SJ, et al. Maternal autoimmune diseases and the risk of autism spectrum disorders in offspring: A systematic review and meta-analysis. Behav Brain Res. 2016;296:61–9.
72. Brimberg L, Sadiq A, Gregersen PK, Diamond B. Brain-reactive IgG correlates with autoimmunity in mothers of a child with an autism spectrum disorder. Mol Psychiatry. 2013;18(11):1171–7.
73. Singer HS, Morris C, Gause C, Pollard M, Zimmerman AW, Pletnikov M. Prenatal exposure to antibodies from mothers of children with autism produces neurobehavioral alterations: A pregnant dam mouse model. J Neuroimmunol. 2009;211(1–2):39–48.
74. Warren RP, Odell JD, Warren WL, Burger RA, Maciulis A, Daniels WW, et al. Strong association of the third hypervariable region of HLA-DR beta 1 with autism. J Neuroimmunol. 1996;67(2):97–102.
75. Chien YL, Wu YY, Chen CH, Gau SS, Huang YS, Chien WH, et al. Association of HLA-DRB1 alleles and neuropsychological function in autism. Psychiatr Genet. 2012;22(1):46–9.
76. Torres AR, Westover JB, Rosenspire AJ. HLA immune function genes in autism. Autism Res Treat. 2012;2012:959073.
77. Abrahams BS, Geschwind DH. Advances in autism genetics: on the threshold of a new neurobiology. Nat Rev Genet. 2008;9(5):341–55.
78. Michel M, Schmidt MJ, Mirnics K. Immune system gene dysregulation in autism and schizophrenia. Dev Neurobiol. 2012;72(10):1277–87.
79. Gaugler T, Klei L, Sanders SJ, Bodea CA, Goldberg AP, Lee AB, et al. Most genetic risk for autism resides with common variation. Nat Genet. 2014;46(8):881–5.

80. Garbett K, Ebert PJ, Mitchell A, Lintas C, Manzi B, Mirnics K, et al. Immune transcriptome alterations in the temporal cortex of subjects with autism. Neurobiol Dis. 2008;30(3):303–11.
81. Masi A, Glozier N, Dale R, Guastella AJ. The immune system, cytokines, and biomarkers in autism spectrum disorder. Neurosci Bull. 2017;33(2):194–204.
82. Matta SM, Hill-Yardin EL, Crack PJ. The influence of neuroinflammation in autism spectrum disorder. Brain Behav Immun. 2019;79:75–90.
83. Saghazadeh A, Ataeinia B, Keynejad K, Abdolalizadeh A, Hirbod-Mobarakeh A, Rezaei N. A meta-analysis of pro-inflammatory cytokines in autism spectrum disorders: Effects of age, gender, and latitude. J Psychiatr Res. 2019;115:90–102.
84. Erbescu A, Papuc SM, Budisteanu M, Arghir A, Neagu M. Re-emerging concepts of immune dysregulation in autism spectrum disorders. Front Psychiatry. 2022;13:1006612.
85. Hughes HK, Moreno RJ, Ashwood P. Innate immune dysfunction and neuroinflammation in autism spectrum disorder (ASD). Brain Behav Immun. 2023;108:245–54.
86. Emanuele E, Orsi P, Boso M, Broglia D, Brondino N, Barale F, et al. Low-grade endotoxemia in patients with severe autism. Neurosci Lett. 2010;471(3):162–5.
87. Ashwood P, Krakowiak P, Hertz-Picciotto I, Hansen R, Pessah IN, Van de Water J. Associations of impaired behaviors with elevated plasma chemokines in autism spectrum disorders. J Neuroimmunol. 2011;232(1–2):196–9.
88. Vargas DL, Nascimbene C, Krishnan C, Zimmerman AW, Pardo CA. Neuroglial activation and neuroinflammation in the brain of patients with autism. Ann Neurol. 2005;57(1):67–81.
89. Morgan JT, Chana G, Pardo CA, Achim C, Semendeferi K, Buckwalter J, et al. Microglial activation and increased microglial density observed in the dorsolateral prefrontal cortex in autism. Biol Psychiatry. 2010;68(4):368–76.
90. Lucchina L, Depino AM. Altered peripheral and central inflammatory responses in a mouse model of autism. Autism Res. 2014;7(2):273–89.
91. Cryan JF, Dinan TG. Mind-altering microorganisms: the impact of the gut microbiota on brain and behaviour. Nat Rev Neurosci. 2012;13(10):701–12.
92. Desbonnet L, Clarke G, Shanahan F, Dinan TG, Cryan JF. Microbiota is essential for social development in the mouse. Mol Psychiatry. 2014;19:146–8.
93. McElhanon BO, McCracken C, Karpen S, Sharp WG. Gastrointestinal symptoms in autism spectrum disorder: a meta-analysis. Pediatrics. 2014;133(5):872–83.
94. Erny D, Hrabě de Angelis AL, Jaitin D, Wieghofer P, Staszewski O, David E, et al. Host microbiota constantly control maturation and function of microglia in the CNS. Nat Neurosci. 2015;18(7):965–77.
95. Goines P, Haapanen L, Boyce R, Duncanson P, Braunschweig D, Delwiche L, et al. Autoantibodies to cerebellum in children with autism associate with behavior. Brain Behav Immun. 2011;25(3):514–23.
96. Wills S, Cabanlit M, Bennett J, Ashwood P, Amaral DG, Van de Water J. Detection of autoantibodies to neural cells of the cerebellum in the plasma of subjects with autism spectrum disorders. Brain Behav Immun. 2009;23(1):64–74.

Chronobiological Aspects of Aging, Neurodegeneration, and Inflammation

Srishti Banerjee, Sandip Das, Nabamita Halder, Sunidhi Bisht, Arunansu Talukdar, and Sandipan Ray

Abstract Our modern-day lifestyle is frequently accompanied by irregular sleep and eating schedules, shift work, and substantial artificial light exposure at night, leading to chronic circadian disruptions. Aberrations in circadian functioning can cause a myriad of physiological and metabolic irregularities, disturbances in the sleep–wake cycle, neurodegeneration, oxidative damage, and a reduced lifespan. Circadian rhythm, sleep, aging, neurodegeneration, and inflammation form a complex interconnected physiological web. The precise circadian timekeeping system deteriorates as we age, leading to decreased adaptability to environmental changes and disruptions in finely tuned circadian regulations of diverse behavioral, physiological, and metabolic processes. Apparent alterations in sleep patterns also emerge with advancing age, potentially elevating the risk of developing neurodegenerative conditions. Disrupted circadian rhythms and poor sleep can further augment systemic inflammation. Reciprocally, chronic inflammation, in turn, can impact sleep quality, creating a complex feedback loop that may deteriorate age-related neurodegenerative processes. Understanding the intricate relationships between circadian disruption, sleep, aging, neurodegeneration, and inflammation is paramount for devising targeted interventions to promote healthy aging and potentially mitigate the risk of neurodegenerative diseases.

Keywords Circadian disruption · Sleep · Aging · Neurodegeneration Inflammation

---

Srishti Banerjee and Sandip Das have equal contribution.

S. Banerjee · S. Das · N. Halder · S. Bisht · S. Ray (✉)
Department of Biotechnology, Indian Institute of Technology Hyderabad, Kandi, Sangareddy, Telangana, India
e-mail: sandipan.ray@bt.iith.ac.in

A. Talukdar
Geriatric Medicine Department, Medical College Hospital Kolkata, Kolkata, West Bengal, India

P. K. Jha, H. Bouaouda (eds.), *Circadian Rhythms, Sleep and Inflammation*, Progress in Inflammation Research 93,
https://doi.org/10.1007/978-3-031-74744-1_8

1 Cross-Talk Among Circadian System, Sleep, Aging, Inflammation, and Neurodegeneration

Circadian rhythm is indispensable to every form of life [1, 2]. The rhythms are controlled by an endogenous timekeeping mechanism, which governs the expression of interconnected genes through transcriptional and translational feedback loops (TTFLs) in a 24-h rhythmic pattern regulating different biological activities [3, 4]. Hence, the circadian system is responsible for the adaptability of life to spatiotemporal cues by synchronizing physiological, behavioral, and metabolic functions.

In mammals, the suprachiasmatic nucleus (SCN) serves as the "master clock" of the body for relaying the external cues received from the environment to the peripheral clocks located in almost every organ and tissue [1, 5]. The 24-h rhythmic oscillations are controlled by a tightly interconnected network of clock genes operating at the molecular level through intricate TTFLs [4]. The molecular circuitry of the mammalian system involves the heterodimerization between transcription factors circadian locomotor output cycle kaput (CLOCK) and brain and muscle ARNT-like 1 (BMAL1) to activate the transcription of Period (*Per*) and Cryptochrome (*Cry*), the other two core clock genes. Subsequently, PER and CRY form a heterodimer to commence the negative feedback loop, repressing the CLOCK-BMAL1-driven transcription [1, 4]. The negative feedback mechanism inhibits the transcription of *Per* and *Cry* genes, marking the completion of an entire cycle and initiating a new one approximately 24 h later.

Similar to biological rhythms, another vital phenomenon of life is aging, which is propelled by sequential manifestations of accumulated degenerative phenomena, including the breakdown of the unequivocal timekeeping machinery [6–9]. Disruption of the circadian clock also predisposes to age-related chronic diseases such as cancers and various neurological, metabolic, inflammatory, and cardiovascular disorders across the lifetime [10–13]. Moreover, as the circadian system controls several processes associated with the absorption, distribution, metabolism, and excretion of medications, circadian-informed therapeutics are emerging steadily for diverse physical and mental diseases [14–16].

Inflammaging is often associated with premature aging, even without a prominent infection, which is also a critical signature of aging [8, 9, 17, 18]. Such inflammation-regulated assaults pervasively affect the normal circadian rhythms, explicitly altering the circadian system in terms of amplitude and regularity of rhythms, abolition of clock gene expression, ultimately reducing the lifespan and leading to premature aging, neurodegeneration, and disrupted sleep–wake cycles [5, 19–22]. Moreover, aging is the most likely demographic predisposing factor toward neurodegenerative disorders [23–25]. Likewise, the inverse correlation between circadian sleep–wake patterns and the pro/anti-inflammatory state of the immune system is evident in its influence on the severity of symptoms and the subsequent immune responses [26]. The causal impact of these disruptions on disease progression needs to be investigated further at the interface of aging, disturbances in sleep patterns, and inflammation-mediated neurodegenerative disorders.

It is becoming increasingly evident that the spectrum of disorders associated with inflammation represents only the surface manifestations of a more profound issue, such as circadian disruption. The fundamental catalyst for this cascade of events resides in the daily behavioral and physiological activities intricately regulated by circadian rhythms. In the relentless engagement of human activities spanning across the diurnal cycle, individuals encounter an environment marked by heightened blue light exposure, irregular meal timings, diminished physical activity, and escalated stress levels [27, 28]. Perturbations in behavioral rhythms subsequently lead to the dysregulation of core clock-controlled gene expression, resulting in a pervasive inflammatory microenvironment and compromised signaling pathways that ultimately predispose individuals to a spectrum of severe disorders (Fig. 1).

This chapter discusses the interrelation among the chronobiological facets of sleep, aging, and neurodegeneration through inflammation. We have additionally perceived the fundamental inflammatory mediators of circadian modulation by delving into the diverse inflammatory pathways intertwined with circadian rhythms concerning the processes of aging, sleep regulation, and neurodegeneration.

2 The Immunity Clock: Unraveling Circadian Control of Aging-Related Inflammation and Neurodegenerative Disorders

Critical components of the immune system show diurnal rhythmic patterns leading to cyclic variations in various host response pathways and immunological networks throughout the day–night cycle [29–31]. Several metabolites and their associated vital metabolic processes also exhibit 24-h rhythmicity in human plasma and blood cells [32, 33]. Aging is a multifaceted degenerative transformation involving myriad disruptions to physiological and metabolic systems, leading to discernible shifts in phenotype [34–36]. Aging affects the core circadian clock components and rhythmicity of numerous clock-controlled genes [6]. The rhythmic expression pattern in clock genes of the SCN extenuates as age advances, consequently impacting neuronal activity [37]. This phenomenon results in diminished amplitudes and substantial desynchronization between cells and tissues, leading to impaired phase shifts. Ultimately, this disruption affects rest–activity patterns and sleep–wake episodes, impairing the onset and progression of age-related diseases [38].

Furthermore, studies involving genetic modifications of core clock genes in mice and *Drosophila* have demonstrated a significant decrease in lifespan [5, 22, 39]. Disruptions in the natural rhythm of circadian patterns have the potential to amplify the aging process. Interestingly, manipulating core clock genes in mammalian organisms has explored the intrinsic connection between aging and the immune system [30]. These experimental knockouts have resulted in modified immune

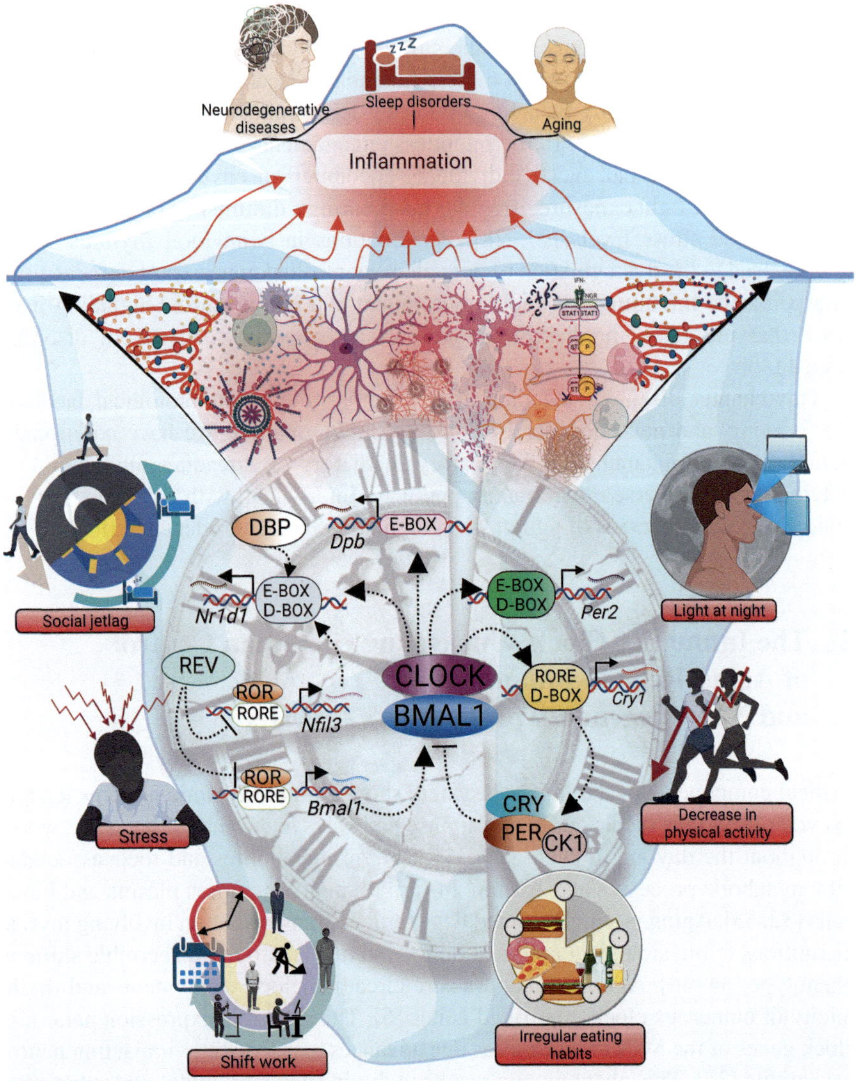

Fig. 1 A schematic depiction of the cross-talk among circadian system, aging, inflammation, and neurodegeneration. Neurodegenerative disorders, sleep disruption, and aging are indispensably associated with inflammation. Circadian rhythms directly regulate our day-to-day behavioral, physiological, and metabolic activities. Since humans in modern society frequently work around the clock, they are exposed to light at night, irregular meal timings, reduced physical exercise, and elevated stress levels, which cumulatively affect behavioral and molecular circadian rhythms and result in the dysregulation of core clock-controlled gene expression. Chronic circadian disruptions can lead to pervasively inflamed microenvironment and dysregulated signaling pathways that substantially enhance the possibility of various physical and mental disorders (created with BioRender.com)

phenotypes, shedding light on the intricate relationship between the circadian clock and immune responses during aging [31].

Disrupted circadian rhythms perturb the regulation of innate and adaptive immune responses, leading to neurodegeneration [40]. Experimental evidence highlights the circadian clock's influence on immune functions specific to immune cell types [41]. The most notable age-related neurological conditions, such as Alzheimer's disease, age-related dementia, and Parkinson's disease, have substantially disrupted circadian rhythms [42]. Occasionally, faltered rhythms even proceed before the onset of clinical symptoms of neurodegeneration [43]. The role of immune cells in eliminating senescent cells is jeopardized with aging as they are affected by immunosenescence, leading to persistent inflammatory conditions [44]. Inflammation and impaired clearance of senescent cells account for a disbalance between the resolution of inflammation and restored tissue homeostasis, leading to age-related pathophysiological conditions [45, 46]. Hence, inflammation is a significant signature of aging [17], connecting aging with circadian disruptions and neurodegeneration.

The circadian regulation over aging, neurodegeneration, and inflammation is exhibited through a multifaceted network of interconnected molecular pathways, including sirtuin-1 (SIRT1) signaling, nuclear factor kappa B (NF-κB) signaling, mammalian/mechanistic target of rapamycin (mTOR) signaling, nuclear factor erythroid 2-related factor (Nrf2) pathway, NOD-, LRR-, and pyrin domain-containing-3 (NLRP3) inflammasome pathway, complement activation, and glucocorticoid signaling pathway (Table 1). These pathways precisely intertwine with the circadian system, culminating in an anti-inflammatory response and serving as positive regulators in attenuating aging and neurodegeneration. SIRT1, which regulates circadian clock parameters through PER2 deacetylation, plays a crucial role in extending lifespan and preventing age-related disorders and neuroprotection through its involvement in sustaining neuronal growth, ensuring genome stability, maintaining metabolic equilibrium, and fostering anti-inflammatory mechanisms [47–49]. SIRT1 induces an anti-inflammatory response through the AMP-activated protein kinase (AMPK) activation pathways targeting the core clock gene *Bmal1*.

Within the NF-κB signaling pathway, the protein NF-κB assumes the role of an inflammatory promoter by stimulating the synthesis of inflammatory markers such as interleukins (IL)-6, IL-8, IL-1β, and tumor necrosis factor α (TNF-α) [55, 56]. Meanwhile, the clock gene *Bmal1*, regulated by retinoid-related orphan receptor α (RORα), facilitates the transcription of inhibitor of nuclear factor kappa B (IκBα), an inhibitor of NF-κB [57–59]. mTOR, a pivotal factor in the aging process and senescence-related disorders, influences cell senescence, hypertrophy, and the development of senescent cells [81]. Recent studies have demonstrated that extending lifespan and alleviating age-related ailments could be achieved by inhibiting mTOR activity by mitigating inflammation and enhancing cellular functionality [82, 83]. The mTOR pathway also regulates synaptic plasticity, memory storage, and neuronal survival [61]. In cellular circadian clock models, decreasing mTOR levels are shown to be associated with elongated circadian periods [63].

Table 1 Major inflammatory pathways connected with circadian functioning, aging, neurodegeneration, and sleep regulation

Sr. no.	Inflammatory pathways	Role in aging and neurodegeneration	Linked circadian clock components	Effects on circadian rhythms	References
1	SIRT1 signaling pathway	SIRT1 promotes prolonged life and delays age-related disorders by regulating AMPK and PGC-1α pathways SIRT-1 is neuroprotective by inhibiting the progression of Ab toxicity by regulating the *FOXO3a* gene Sleep deprivation results in reduced SIRT1 expression and, blocked anti-inflammatory actions, and reduced delta waves during NREM sleep	BMAL1, CLOCK, PER2	Reduced SIRT1 levels coincide with decreased BMAL1-CLOCK and PER2 expression, leading to an extended period length and impaired synchrony to the light schedule	[47–52]
2	NF-κB signaling pathway	The key player in initiating senescence-associated secretory phenotype (SASP) is NF-κB, which fosters the production of inflammatory markers such as IL-6, IL-8, IL-1β, and TNF-α NF-κB maintains immune and circadian homeostasis. On overstimulation due to inflammation, it releases proinflammatory cytokines in microglia and enzymes like BACE-1, which facilitates Alzheimer's progression Sleep loss enhances the production of proinflammatory cytokines and chemokines, which are regulated by genes activated through the I-κB-dependent NF-κB activation pathway	CLOCK, RORα, BMAL1	BMAL1 controls RORα, which triggers the transcription of IκBα, an NF-κB pathway inhibitor. Additionally, CLOCK promotes NF-κB activation, resulting in excessive production of inflammatory cytokines	[53–60]

Sr. no.	Inflammatory pathways	Role in aging and neurodegeneration	Linked circadian clock components	Effects on circadian rhythms	References
3	mTOR pathway	Impaired mTORC1 activity dysregulates proteostasis, contributing to aging and age-related disorders like Alzheimer's Sleep deprivation weakens hippocampal mTORC1, reducing protein synthesis and impairing memory.	CRY1, BMAL1, CLOCK	mTOR activation induces CRY1, BMAL1, CLOCK protein expression. In cellular clock models, a reduction in the mTOR levels leads to longer circadian periods	[61–65]
4	Nrf2 pathway	Activated NRF2 prolongs and ameliorates age-related characteristics through reactive oxygen species regulation Sleep loss results in elevated proinflammatory cytokines production and reduces Keap1-Nrf2 signaling, cumulatively triggering neuroinflammation	BMAL1, CLOCK	Clock components CLOCK and BMAL1 interact with the Nrf2/ARE promoter via the E-box element, thereby controlling the periodic activation of the Nrf2 transcription factor	[66–69]

(continued)

Table 1 (continued)

Sr. no.	Inflammatory pathways	Role in aging and neurodegeneration	Linked circadian clock components	Effects on circadian rhythms	References
5	NLRP3 inflammasome pathway	Suppression of the NLRP3 inflammasome complex prolongs lifespan by alleviating age-related degeneration The assembly of the NLRP3 inflammasome leads to caspase-1 activation, triggering the release of proinflammatory cytokines IL-1β and IL-18 in sleep	Rev-erbα	The NLRP3 inflammasome undergoes daily fluctuations under the control of Rev-erbα with peak expression during the active phase and lowest expression during the inactive phase (similar to Rev-erbα expression). Also, Rev-erba plays an anti-inflammatory and neuroprotective role by inhibiting the microglial activation	[60, 70–73]
6	Complement activation system	Elevated levels of C3 and C4 exhibit a negative correlation with lifespan, with increased levels of C3 in the later part of life Complement proteins C1q, C3, and C4 are found at elevated levels in the brains of neurodegenerative diseases, along with amyloid-beta (Aβ) plaques Sleep deprivation results in neuroinflammation via complement (C3a-C3aR and C5a-C5aR) activation	CLOCK/ BMAL1 and REV-ERB- α/β	The CLOCK/BMAL1 complex triggers the expression of *Per* (1, 2, 3), *Cry* (1, 2); *Rev-erb- α/β* leading to elevated levels of circadian regulatory proteins CRY1/CRY2 repressors, CD59, CRRY, DAF and HO-1. Consequently, this leads to decreased levels of C3, C5a, and inhibits membrane attack complex formation	[74–76]
7	Glucocorticoid signaling pathway	Amplified levels of glucocorticoids hinder hippocampal functions and impair spatial memory in aging and age-related diseases Sleeping out-of-phase results in increased expression of MAPK11 elevated levels of inflammatory cytokines	PER1, PER2, REV-ERBα, and NFIL3	Glucocorticoids regulate clock components PER (1,2), REV-ERBα, and NFIL3 via GRE motifs. The central nervous system controls diurnal GC production, coordinating the circadian rhythm in peripheral organs	[77–80]

During the aging process, a decline in the activity of the Kelch-like ECH-associated protein 1-NFE2-related factor 2 (KEAP1-NRF2) system is observed due to an increase in senescent cells in specific tissues [84]. Nrf2, a crucial transcription factor in cellular defense against oxidative stress, helps to maintain redox balance in microglia [69]. Notably, the core clock protein BMAL1 interacts with the Nrf2/ARE promoter via the E-box element, thereby governing the rhythmic activation of the Nrf2 transcription factor [67, 68, 85].

In aged mice lacking NLRP3, there is an increase in naive T cells and a decrease in effector-memory cells, indicating the involvement of the NLRP3 inflammasome in thymic involution and immunosenescence [134]. Suppression of the NLRP3 inflammasome, improves age-related degenerative changes, including cognitive decline [70, 73]. The NLRP3 inflammasome experiences daily variations, regulated by Rev-erbα, with its highest expression observed during the active phase and lowest expression occurring during the inactive phase [73].

In the aging process, inflammation plays a significant role, and the activation of the complement system via C3 hydrolysis is a key mechanism [76]. Elevated levels of complement components C3 and C4 have been associated with aging, and higher C3 levels are linked to reduced longevity, suggesting potentially harmful effects on lifespan [76]. The transcriptional activation of nuclear receptor subfamily 1 group d member 1 (REV-ERB) by CLOCK/BMAL1 leads to increased levels of circadian-regulated proteins, including protectin (CD59), complement component receptor 1-like protein (CRRY), decay accelerating factor (DAF), and antioxidant factors like hemoxygenase1 (HO-1) [74]. This, in turn, decreases levels of C3 and C5a and inhibits membrane attack complex (MAC) formation, thereby regulating complement activity [74].

Prolonged exposure to elevated glucocorticoid (GCs) levels negatively impacts the hippocampus's functioning [80]. There is a connection between the age-related rise in blood GC levels and a decline in spatial memory abilities [80]. Also, through GRE motifs, glucocorticoids modulate clock elements PER (1,2), REV-ERBα, and NFIL3 [79].

Aging and chronic inflammation are intricately connected to elevated Toll-like receptor (TLR) expression [86–88]. In older individuals, TLR5 and TLR4 expression is upregulated, which correlates with increased cytokine production and impaired muscle insulin [86–88]. Several signaling pathways contribute to the changes in TLR function during aging, such as reduced phosphoinositide 3-kinases (PI3K) activity and impaired downregulation of signal transducer and activator of transcription (STAT1) phosphorylation, leading to enhanced TLR-dependent cytokine production [86–88]. Mutations in the *Per2* gene, a crucial component of the molecular clock, can impact the expression and function of TLR9 in specific immune cells [88].

The Janus kinase/signal transducers and activators of transcription (JAK/STAT) pathway, a pivotal regulator of cytokine production, orchestrates critical cellular processes, including hematopoiesis, inflammation, tissue regeneration, and immune functionality [89]. Since acquiring the senescence-associated secretory phenotype (SASP) phenotype characterizes aging, the JAK pathway's enhanced activity is observed in senescent cells [90]. Consequently, targeted inhibition of senescent cells via modulation of the JAK pathway mitigates SASP expression and ameliorates age-associated dysfunctions [90]. Therefore, this pathway emerges as a potential marker of frailty [91]. Genetic manipulation of JAK/STAT through *miR-279* uncovered the role of the JAK/STAT pathway in behavioral rhythms in *Drosophila* [92].

3 Sleep Deprivation-Induced Activation of Inflammatory Pathways

The homeostatically regulated behavioral phenomenon "sleep" is harmoniously tangled with the body clock in the intricate tapestry of life's rhythms. Light plays a pivotal role as an external cue in synchronizing the body clock [93]. Mistimed exposure to light–dark cycles, such as social jet lag, shift work, and late-night activities, can lead to the desynchronization of the internal clock [93], leading to sleep disruption [94]. Sleep restriction studies indicate cell-specific control of transcriptional, translational, and post-translational responses by sleep need [95]. It has been reported that sleep disruptions are commonly observed among the elderly populations, marked by earlier initiation of sleep, fragmented sleep episodes, and substantially reduced quality of sleep [96]. Additionally, this bidirectionally regulated association between sleep and the circadian clock extensively affects aging and age-related neurodegenerative disorders such as Alzheimer's, Parkinson's, and Huntington's diseases [97]. Individuals with age-related neurodegenerative disorders have more profound dysfunctional circadian rhythms, comprising of fragmented sleep (phase delay in the sleep–wake cycle), dampened amplitude of oscillations, shifts in the phase of hormonal and temperature rhythms [98].

A disrupted sleep–wake cycle indicates a malfunctioning internal clock leading to an insidious proinflammatory condition [60]. Melatonin and cortisol exhibit opposite-phased oscillations, fostering periods of sleep and wakefulness in sync with the light–dark cycle. Cortisol and melatonin are anti-inflammatory agents that suppress inflammatory cytokine production [99, 100]. Systemic inflammation indicators, including C-reactive protein (CRP), IL-6, and fibrinogen, have been correlated with disrupted sleep patterns, as elevated levels of these markers are associated with lower sleep quality [101]. Sleep is connected with inflammation through circadian regulation in a complex network of molecular pathways. Several signaling pathways like SIRT1 signaling, NF-κβ signaling, mTOR, Nrf2 signaling, NLRP3 signaling, complement activation, glucocorticoid signaling pathway, and their associated molecules play a crucial role in sleep disturbance-induced inflammation process (Table 1). Moreover, several pharmacological modulators of mammalian kinases alter circadian parameters [102, 103].

Sleep deprivation causes elevated production of proinflammatory components IL-1β and TNF-α and results in CLOCK–BMAL1 heterodimer-regulated, COX2-induced prostaglandin-stimulated acute inflammation [104–109].

The CLOCK-controlled NF-κB gene product undergoes diurnal oscillations, with its peak occurring during periods of elevated sleep pressure [59, 60]. Consequently, sleep deprivation-induced wakefulness increases proinflammatory NF-κB production [109–111] and leads to the transcription of proinflammatory molecules [112, 113]. The nuclear receptor REV-ERBα, which links the circadian clock to energy homeostasis, orchestrates the diurnal regulation of the crucial elements of the NLRP3 inflammasome pathway, such as caspase-1, IL-1β, and IL-18 [73]. Consequently, this exerts a direct influence on an individual's sleep propensity. CLOCK, BMAL1, and *Per2* govern the TLR signaling pathway, triggering inflammatory responses through the heightened release of proinflammatory cytokines during wakefulness [88, 114, 115]. SIRT1 autonomously operates the clock core circuitry [47, 52], and its impact extends to sleep-related behavioral aspects, such as the preservation of delta waves during non-rapid eye-movement (NREM) sleep. Furthermore, SIRT1 significantly regulates sleep–wake neurotransmitters and somnogens [51, 116].

Sleep deprivation with stress activates the JAK-STAT3 pathway, which, in turn, activates the proinflammatory cytokines like IL-6 and TNF-α, leading to an acute inflammatory response [117]. Additionally, IL-4 positively regulates *Per2* expression at the mRNA level through the JAK/STAT3 signaling pathway [118]. The scarcity of sleep inhibits the hippocampus mTORC1 pathway, which reduces protein synthesis and impairs memory [65]. Impaired memory may be restored by reinstating mTORC1 signaling [119]. Reduction in mTOR levels results in the expression of CRY1 with extended circadian periods [63].

Sleep deprivation is associated with complement activation, specifically C3 [75]. Notably, during sleep loss, there is an upregulation of C3aR and C5aR receptor expression, indicating their potential involvement in mediating the effects of cytokine-induced immune responses [75]. The CLOCK/BMAL1 complex regulates the activation of REV-ERB α/β, suppressing C3 and C5a production and inhibiting MAC formation [74]. Moreover, BMAL1 plays a pivotal role in governing the rhythmic activation of the Nrf2 transcription factor by binding to the Nrf2/ARE promoter via the E-box element [66, 85]. Sleep loss disrupts Nrf2 signaling, leading to an increase in the production of proinflammatory cytokines IL-1, IL-6, TNF-α, COX-2, and inducible nitric oxide synthase (iNOS) [66, 120, 68].

The glucocorticoid signaling pathway regulates the expressions of PER1, PER2, REV-ERBα, and nuclear factor interleukin 3 regulated (NFIL3) [78]. Misaligned rhythms from sleep disturbances disrupt the transcriptional repressor, nuclear receptor corepressor 1 (NCOR1), protein kinase mitogen-activated protein kinase 1 (MAPK1), and the heat shock glucocorticoid chaperone proteins, heat shock protein 90-alpha (HSP90AA1) and heat shock protein family A (Hsp70) member 1A HSPA1A cumulatively leading to metabolic disorder and inflammation [77]. These studies cumulatively emphasize the critical involvement of sleep loss and misaligned circadian rhythms in promoting inflammation.

4 Conclusive Remarks and Translational Outlook

The complex molecular mechanisms promoting inflammation indicate a focal point of convergence among aging, sleep disorders, and neurodegenerative disorders (Fig. 2). With aging, various anti-inflammatory regulators like a disintegrin and metalloproteinase 10 (ADAM10), Nrf2, and endothelial nitric oxide synthase

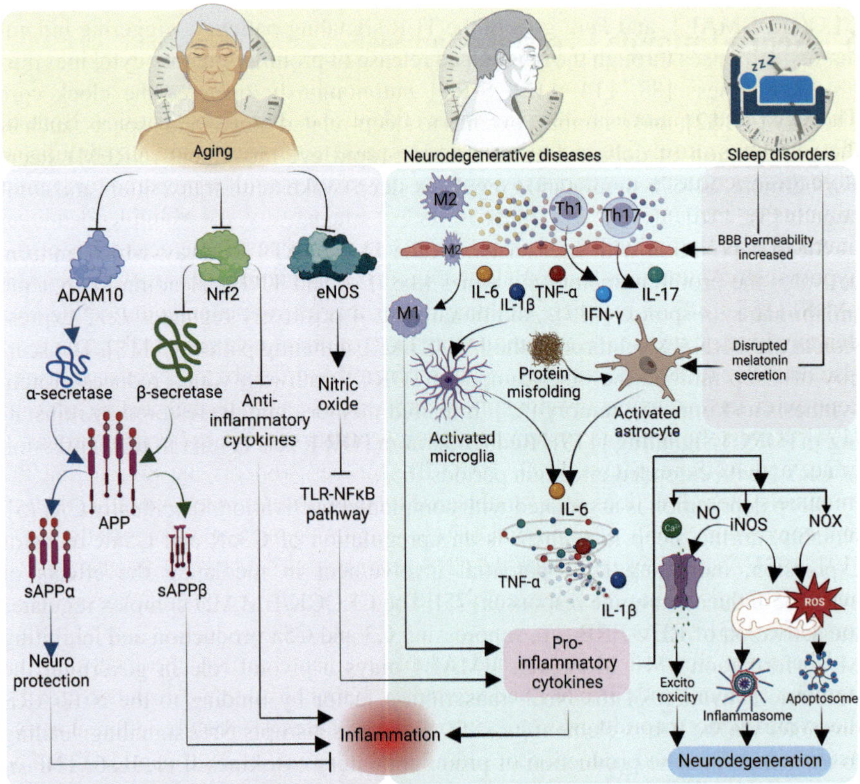

Fig. 2 Molecular facets of inflammation, a binding thread between aging, neurodegeneration, and sleep disruption. Aging disrupts crucial anti-inflammatory regulators in the cerebral vasculature, down-regulating ADAM10, Nrf2, and eNOS expression. ADAM10 cleaves APP to produce neuroprotective soluble APPα, while Nrf2 negatively regulates APP cleavage, contributing to amyloid-β deposition and neuroinflammation. Reduced eNOS levels enhance oxidative stress, activating the TLR-NF-κB pathway and triggering proinflammatory cytokine secretion, leading to a feedback loop aggravating neuroinflammation. Aging also induces cellular and molecular changes in the brain, with increased proinflammatory cytokines and protein accumulation, disrupting the blood-brain barrier and driving low-grade brain inflammation. Activated microglia and astrocytes produce proinflammatory cytokines, while macrophages shift to a proinflammatory phenotype, further fueling neuroinflammation. Proinflammatory cytokines activate the ROS signaling pathway, causing mitochondrial dysfunction, inflammasome, apoptosome generation, and excitotoxicity, leading to neurodegeneration. Sleep loss and disrupted melatonin secretion increase blood-brain barrier permeability, further exacerbating inflammation (created with BioRender.com)

(eNOS) are downregulated and lead to the activation of proinflammatory cytokines via the TLR-NF-κB [121–126]. The aging process also increases the blood-brain barrier's permeability, allowing immune cells to infiltrate the brain parenchyma, affecting the microglia phenotype and macrophages [122, 127].

The brain parenchyma also experiences a cytokine storm characterized by elevated levels of proinflammatory cytokines, including IL-1, IL-6, and TNF [123, 128–130]. The accumulation of proinflammatory cytokines ultimately leads to excitotoxicity and neurodegeneration by activating the reactive oxygen species (ROS) signaling pathway and generating inflammasome and apoptosome complexes [123, 131]. Furthermore, sleep disruption induces disrupted melatonin secretion, increases the permeability of the blood-brain barrier, and results in inflammation [132]. Also, the significant inflammation regulators closely interact with the circadian system (Table 1).

With the advent of chronic inflammation models, it is well under investigation to ameliorate premature aging by reducing the pervasive nature of inflammation [133]. Given the restricted efficacy of existing treatments for several neurological disorders, it is of rising interest to regulate inflammatory signaling pathways in neurodegeneration as a prospective therapy. In conclusion, the significant impact of inflammation on aging, sleep, and neurodegeneration makes it an up-and-coming candidate for therapeutic intervention. Moreover, an in-depth investigation of chronobiological aspects of neuroinflammation can have essential implications for early diagnosis and therapeutics of neurodegenerative diseases.

Acknowledgments SR acknowledges funding from the Indian Council of Medical Research (ICMR/BMS/Adhoc/184/2022-23) and the Science and Engineering Research Board, Department of Science and Technology, Government of India (SRG/2021/000671). SB is supported by the Prime Minister's Research Fellows (PMRF) Scheme, Government of India (PMRF 2002213). SD is supported by a Council of Scientific and Industrial Research-University Grants Commission fellowship (UGC-WB10602148). All the figures in this book chapter are prepared using BioRender.com.

Compliance with Ethical Standards This article does not contain any studies with human participants or animals performed by any of the authors

Conflict of Interest The authors declare no conflict of interest, financial or otherwise.

References

1. Bell-Pedersen D, Cassone VM, Earnest DJ, Golden SS, Hardin PE, Thomas TL, et al. Circadian rhythms from multiple oscillators: lessons from diverse organisms. Nat Rev Genet. 2005;6(7):544–56.
2. Ray S, Reddy AB. Cross-talk between circadian clocks, sleep-wake cycles, and metabolic networks: Dispelling the darkness. Bioessays. 2016;38(4):394–405.
3. Patke A, Young MW, Axelrod S. Molecular mechanisms and physiological importance of circadian rhythms. Nat Rev Mol Cell Biol. 2020;21(2):67–84.

4. Takahashi JS. Transcriptional architecture of the mammalian circadian clock. Nat Rev Genet. 2017;18(3):164–79.
5. Dubrovsky YV, Samsa WE, Kondratov RV. Deficiency of circadian protein CLOCK reduces lifespan and increases age-related cataract development in mice. Aging (Albany NY). 2010;2(12):936–44.
6. Duffy JF, Zitting K-M, Chinoy ED. Aging and circadian rhythms. Sleep Med Clin. 2015;10(4):423–34.
7. López-Otín C, Blasco MA, Partridge L, Serrano M, Kroemer G. Hallmarks of aging: An expanding universe. Cell. 2023;186(2):243–78.
8. López-Otín C, Blasco MA, Partridge L, Serrano M, Kroemer G. The hallmarks of aging. Cell. 2013;153(6):1194–217.
9. Schmauck-Medina T, Molière A, Lautrup S, Zhang J, Chlopicki S, Madsen HB, et al. New hallmarks of ageing: a 2022 Copenhagen ageing meeting summary. Aging (Albany NY). 2022;14(16):6829–39.
10. Chellappa SL, Vujovic N, Williams JS, Scheer FAJL. Impact of circadian disruption on cardiovascular function and disease. Trends Endocrinol Metab. 2019;30(10):767–79.
11. Nassan M, Videnovic A. Circadian rhythms in neurodegenerative disorders. Nat Rev Neurol. 2022;18(1):7–24.
12. Sulli G, Manoogian ENC, Taub PR, Panda S. Training the circadian clock, clocking the drugs, and drugging the clock to prevent, manage, and treat chronic diseases. Trends Pharmacol Sci. 2018;39(9):812–27.
13. Verlande A, Masri S. Circadian clocks and cancer: timekeeping governs cellular metabolism. Trends Endocrinol Metab. 2019;30(7):445–58.
14. Bhatnagar A, Murray G, Ray S. Circadian biology to advance therapeutics for mood disorders. Trends Pharmacol Sci. 2023;44(10):689–704.
15. Chang YC, Kim JY. Therapeutic implications of circadian clocks in neurodegenerative diseases. J Neurosci Res. 2020;98(6):1095–113.
16. Puppala A, Rankawat S, Ray S. Circadian timekeeping in anticancer therapeutics: an emerging vista of chronopharmacology research. Curr Drug Metab. 2021;22(13):998–1008.
17. Campisi J, Kapahi P, Lithgow GJ, Melov S, Newman JC, Verdin E. From discoveries in ageing research to therapeutics for healthy ageing. Nature. 2019;571(7764):183–92.
18. Sanada F, Taniyama Y, Muratsu J, Otsu R, Shimizu H, Rakugi H, et al. Source of chronic inflammation in aging. Front Cardiovasc Med. 2018;5:12.
19. Banerjee S, Ray S. Circadian medicine for aging attenuation and sleep disorders: prospects and challenges. Prog Neurobiol. 2022;220:102387.
20. Nakamura TJ, Nakamura W, Yamazaki S, Kudo T, Cutler T, Colwell CS, et al. Age-related decline in circadian output. J Neurosci. 2011;31(28):10201–5.
21. Sato S, Solanas G, Peixoto FO, Bee L, Symeonidi A, Schmidt MS, et al. Circadian reprogramming in the liver identifies metabolic pathways of aging. Cell. 2017;170(4):664–677.e11.
22. Vaccaro A, Issa A-R, Seugnet L, Birman S, Klarsfeld A. Drosophila clock is required in brain pacemaker neurons to prevent premature locomotor aging independently of its circadian function. PLoS Genet. 2017;13(1):e1006507.
23. Armstrong RA. Risk factors for Alzheimer's disease. Folia Neuropathol. 2019;57(2):87–105.
24. Ravindranath V, Sundarakumar JS. Changing demography and the challenge of dementia in India. Nat Rev Neurol. 2021;17(12):747–58.
25. van der Flier WM, Scheltens P. Epidemiology and risk factors of dementia. J Neurol Neurosurg Psychiatry. 2005;76(Suppl 5):v2–7.
26. Comas M, Gordon CJ, Oliver BG, Stow NW, King G, Sharma P, et al. A circadian based inflammatory response—implications for respiratory disease and treatment. Sleep Sci Pract. 2017;1(1):18.
27. Lockley SW, Brainard GC, Czeisler CA. High sensitivity of the human circadian melatonin rhythm to resetting by short wavelength light. J Clin Endocrinol Metab. 2003;88(9):4502–5.

28. Salgado-Delgado RC, Saderi N, Basualdo MDC, Guerrero-Vargas NN, Escobar C, Buijs RM. Shift work or food intake during the rest phase promotes metabolic disruption and desynchrony of liver genes in male rats. PLoS One. 2013;8(4):e60052.
29. Rankawat S, Kundal K, Chakraborty S, Kumar R, Ray S. A comprehensive rhythmicity analysis of host proteins and immune factors involved in malaria pathogenesis to decipher the importance of host circadian clock in malaria. Front Immunol. 2023;14:1210299.
30. Scheiermann C, Gibbs J, Ince L, Loudon A. Clocking in to immunity. Nat Rev Immunol. 2018;18(7):423–37.
31. Wang C, Lutes LK, Barnoud C, Scheiermann C. The circadian immune system. Sci Immunol. 2022;7(72):eabm2465.
32. Ch R, Rey G, Ray S, Jha PK, Driscoll PC, Dos Santos MS, et al. Rhythmic glucose metabolism regulates the redox circadian clockwork in human red blood cells. Nat Commun. 2021;12(1):377.
33. Dallmann R, Viola AU, Tarokh L, Cajochen C, Brown SA. The human circadian metabolome. Proc Natl Acad Sci USA. 2012;109(7):2625–9.
34. Dziechciaż M, Filip R. Biological psychological and social determinants of old age: bio-psycho-social aspects of human aging. Ann Agric Environ Med. 2014;21(4):835–8.
35. Johnson FB, Sinclair DA, Guarente L. Molecular biology of aging. Cell. 1999;96(2):291–302.
36. Murman DL. The impact of age on cognition. Semin Hear. 2015;36(3):111–21.
37. Yamazaki S, Straume M, Tei H, Sakaki Y, Menaker M, Block GD. Effects of aging on central and peripheral mammalian clocks. Proc Natl Acad Sci USA. 2002;99(16):10801–6.
38. De Nobrega AK, Lyons LC. Aging and the clock: Perspective from flies to humans. Eur J Neurosci. 2020;51(1):454–81.
39. Kondratov RV, Kondratova AA, Gorbacheva VY, Vykhovanets OV, Antoch MP. Early aging and age-related pathologies in mice deficient in BMAL1, the core component of the circadian clock. Genes Dev. 2006;20(14):1868–73.
40. Wang X-L, Li L. Circadian clock regulates inflammation and the development of neurodegeneration. Front Cell Infect Microbiol. 2021;11:696554.
41. Reglero-Real N, Rolas L, Nourshargh S. Leukocyte trafficking: time to take time seriously. Immunity. 2019;50(2):273–5.
42. Leng Y, Musiek ES, Hu K, Cappuccio FP, Yaffe K. Association between circadian rhythms and neurodegenerative diseases. Lancet Neurol. 2019;18(3):307–18.
43. Tranah GJ, Blackwell T, Stone KL, Ancoli-Israel S, Paudel ML, Ensrud KE, et al. Circadian activity rhythms and risk of incident dementia and mild cognitive impairment in older women. Ann Neurol. 2011;70(5):722–32.
44. Li X, Li C, Zhang W, Wang Y, Qian P, Huang H. Inflammation and aging: signaling pathways and intervention therapies. Signal Transduct Target Ther. 2023;8(1):239.
45. Ferrucci L, Fabbri E. Inflammageing: chronic inflammation in ageing, cardiovascular disease, and frailty. Nat Rev Cardiol. 2018;15(9):505–22.
46. Sendama W. The effect of ageing on the resolution of inflammation. Ageing Res Rev. 2020;57:101000.
47. Asher G, Gatfield D, Stratmann M, Reinke H, Dibner C, Kreppel F, et al. SIRT1 regulates circadian clock gene expression through PER2 deacetylation. Cell. 2008;134(2):317–28.
48. Chang H-C, Guarente L. SIRT1 mediates central circadian control in the SCN by a mechanism that decays with aging. Cell. 2013;153(7):1448–60.
49. Mishra P, Mittal AK, Kalonia H, Madan S, Ghosh S, Sinha JK, et al. SIRT1 promotes neuronal fortification in neurodegenerative diseases through attenuation of pathological hallmarks and enhancement of cellular lifespan. Curr Neuropharmacol. 2021;19(7):1019–37.
50. Hardeland R. Aging, melatonin, and the pro- and anti-inflammatory networks. Int J Mol Sci. 2019;20(5):1223.
51. Mehramiz M, Porter T, Laws SM, Rainey-Smith SR. Sleep, Sirtuin 1 and Alzheimer's disease: A review. Aging Brain. 2022;2:100050.

52. Nakahata Y, Kaluzova M, Grimaldi B, Sahar S, Hirayama J, Chen D, et al. The NAD+-dependent deacetylase SIRT1 modulates CLOCK-mediated chromatin remodeling and circadian control. Cell. 2008;134(2):329–40.
53. Chien Y, Scuoppo C, Wang X, Fang X, Balgley B, Bolden JE, et al. Control of the senescence-associated secretory phenotype by NF-κB promotes senescence and enhances chemosensitivity. Genes Dev. 2011;25(20):2125–36.
54. Irwin MR, Wang M, Ribeiro D, Cho HJ, Olmstead R, Breen EC, et al. Sleep loss activates cellular inflammatory signaling. Biol Psychiatry. 2008;64(6):538–40.
55. Jones SV, Kounatidis I. Nuclear factor-kappa B and Alzheimer disease, unifying genetic and environmental risk factors from cell to humans. Front Immunol. 2017;8:1805.
56. Kim MK, Chung SW, Kim DH, Kim JM, Lee EK, Kim JY, et al. Modulation of age-related NF-kappaB activation by dietary zingerone via MAPK pathway. Exp Gerontol. 2010;45(6):419–26.
57. Ni J, Wu Z, Meng J, Saito T, Saido TC, Qing H, et al. An impaired intrinsic microglial clock system induces neuroinflammatory alterations in the early stage of amyloid precursor protein knock-in mouse brain. J Neuroinflamm. 2019;16(1):173.
58. Shen Y, Endale M, Wang W, Morris AR, Francey LJ, Harold RL, et al. NF-κB modifies the mammalian circadian clock through interaction with the core clock protein BMAL1. PLoS Genet. 2021;17(11):e1009933.
59. Spengler ML, Kuropatwinski KK, Comas M, Gasparian AV, Fedtsova N, Gleiberman AS, et al. Core circadian protein CLOCK is a positive regulator of NF-κB-mediated transcription. Proc Natl Acad Sci USA. 2012;109(37):E2457–65.
60. Zielinski MR, Gibbons AJ. Neuroinflammation, sleep, and circadian rhythms. Front Cell Infect Microbiol. 2022;12:853096.
61. Bockaert J, Marin P. mTOR in brain physiology and pathologies. Physiol Rev. 2015;95(4):1157–87.
62. Liu W, Zhao D, Wu X, Yue F, Yang H, Hu K. Rapamycin ameliorates chronic intermittent hypoxia and sleep deprivation-induced renal damage via the mammalian target of rapamycin (mTOR)/NOD-like receptor protein 3 (NLRP3) signaling pathway. Bioengineered. 2022;13(3):5537–50.
63. Ramanathan C, Kathale ND, Liu D, Lee C, Freeman DA, Hogenesch JB, et al. mTOR signaling regulates central and peripheral circadian clock function. PLoS Genet. 2018;14(5):e1007369.
64. Tramutola A, Triplett JC, Di Domenico F, Niedowicz DM, Murphy MP, Coccia R, et al. Alteration of mTOR signaling occurs early in the progression of Alzheimer disease (AD): analysis of brain from subjects with pre-clinical AD, amnestic mild cognitive impairment and late-stage AD. J Neurochem. 2015;133(5):739–49.
65. Tudor JC, Davis EJ, Peixoto L, Wimmer ME, van Tilborg E, Park AJ, et al. Sleep deprivation impairs memory by attenuating mTORC1-dependent protein synthesis. Sci Signal. 2016;9(425):ra41.
66. Ahmed A, Misrani A, Tabassum S, Yang L, Long C. Minocycline inhibits sleep deprivation-induced aberrant microglial activation and Keap1-Nrf2 expression in mouse hippocampus. Brain Res Bull. 2021;174:41–52.
67. Chhunchha B, Kubo E, Singh DP. Clock protein Bmal1 and Nrf2 cooperatively control aging or oxidative response and redox homeostasis by regulating rhythmic expression of Prdx6. Cells. 2020;9(8):1861.
68. Saha S, Buttari B, Profumo E, Tucci P, Saso L. A perspective on Nrf2 signaling pathway for neuroinflammation: a potential therapeutic target in Alzheimer's and Parkinson's diseases. Front Cell Neurosci. 2021;15:787258.
69. Villavicencio Tejo F, Quintanilla RA. Contribution of the Nrf2 pathway on oxidative damage and mitochondrial failure in Parkinson and Alzheimer's disease. Antioxidants (Basel). 2021;10(7):1069.

70. Blevins HM, Xu Y, Biby S, Zhang S. The NLRP3 inflammasome pathway: a review of mechanisms and inhibitors for the treatment of inflammatory diseases. Front Aging Neurosci. 2022;14:879021.
71. Kou L, Chi X, Sun Y, Han C, Wan F, Hu J, et al. The circadian clock protein Rev-erbα provides neuroprotection and attenuates neuroinflammation against Parkinson's disease via the microglial NLRP3 inflammasome. J Neuroinflamm. 2022;19(1):133.
72. Pandey A, Shen C, Feng S, Man SM. Cell biology of inflammasome activation. Trends Cell Biol. 2021;31(11):924–39.
73. Pourcet B, Duez H. Circadian control of inflammasome pathways: implications for circadian medicine. Front Immunol. 2020;11:1630.
74. Shivshankar P, Fekry B, Eckel-Mahan K, Wetsel RA. Circadian clock and complement immune system-complementary control of physiology and pathology? Front Cell Infect Microbiol. 2020;10:418.
75. Wadhwa M, Prabhakar A, Anand JP, Ray K, Prasad D, Kumar B, et al. Complement activation sustains neuroinflammation and deteriorates adult neurogenesis and spatial memory impairment in rat hippocampus following sleep deprivation. Brain Behav Immun. 2019;82:129–44.
76. Zheng R, Zhang Y, Zhang K, Yuan Y, Jia S, Liu J. The complement system, aging, and aging-related diseases. Int J Mol Sci. 2022;23(15):8689.
77. Archer SN, Möller-Levet CS, Laing EE, Dijk D-J. Mistimed sleep and waking activity in humans disrupts glucocorticoid signalling transcripts and SP1, but not plasma cortisol rhythms. Front Physiol. 2022;13:946444.
78. Shimba A, Ikuta K. Glucocorticoids regulate circadian rhythm of innate and adaptive immunity. Front Immunol. 2020;11:2143.
79. So AY-L, Bernal TU, Pillsbury ML, Yamamoto KR, Feldman BJ. Glucocorticoid regulation of the circadian clock modulates glucose homeostasis. Proc Natl Acad Sci USA. 2009;106(41):17582–7.
80. Yau JLW, Seckl JR. Local amplification of glucocorticoids in the aging brain and impaired spatial memory. Front Aging Neurosci. 2012;4:24.
81. Chrienova Z, Nepovimova E, Kuca K. The role of mTOR in age-related diseases. J Enzyme Inhib Med Chem. 2021;36(1):1679–93.
82. Johnson SC, Rabinovitch PS, Kaeberlein M. mTOR is a key modulator of ageing and age-related disease. Nature. 2013;493(7432):338–45.
83. Mannick JB, Lamming DW. Targeting the biology of aging with mTOR inhibitors. Nat Aging. 2023;3(6):642–60.
84. Matsumaru D, Motohashi H. The KEAP1-NRF2 system in healthy aging and longevity. Antioxidants (Basel). 2021;10(12):1929.
85. Bevinakoppamath S, Ramachandra SC, Yadav AK, Basavaraj V, Vishwanath P, Prashant A. Understanding the emerging link between circadian rhythm, Nrf2 pathway, and breast cancer to overcome drug resistance. Front Pharmacol. 2021;12:719631.
86. Chung HY, Kim DH, Lee EK, Chung KW, Chung S, Lee B, et al. Redefining chronic inflammation in aging and age-related diseases: proposal of the senoinflammation concept. Aging Dis. 2019;10(2):367–82.
87. Shaw AC, Panda A, Joshi SR, Qian F, Allore HG, Montgomery RR. Dysregulation of human Toll-like receptor function in aging. Ageing Res Rev. 2011;10(3):346–53.
88. Silver AC, Arjona A, Walker WE, Fikrig E. The circadian clock controls toll-like receptor 9-mediated innate and adaptive immunity. Immunity. 2012;36(2):251–61.
89. Hu X, Li J, Fu M, Zhao X, Wang W. The JAK/STAT signaling pathway: from bench to clinic. Signal Transduct Target Ther. 2021;6(1):402.
90. Xu M, Tchkonia T, Kirkland JL. Perspective: Targeting the JAK/STAT pathway to fight age-related dysfunction. Pharmacol Res. 2016;111:152–4.
91. Samson LD, Engelfriet P, Verschuren WMM, Picavet HSJ, Ferreira JA, de Zeeuw-Brouwer M-L, et al. Impaired JAK-STAT pathway signaling in leukocytes of the frail elderly. Immun Ageing. 2022;19(1):5.

92. Luo W, Sehgal A. Regulation of circadian behavioral output via a MicroRNA-JAK/STAT circuit. Cell. 2012;148(4):765–79.
93. Roenneberg T, Merrow M. The circadian clock and human health. Curr Biol. 2016;26(10):R432–43.
94. Abbott SM, Reid KJ, Zee PC. Circadian rhythm sleep-wake disorders. Psychiatr Clin North Am. 2015;38(4):805–23.
95. Jha PK, Valekunja UK, Ray S, Nollet M, Reddy AB. Single-cell transcriptomics and cell-specific proteomics reveals molecular signatures of sleep. Commun Biol. 2022;5(1):846.
96. Dijk DJ, Duffy JF, Czeisler CA. Age-related increase in awakenings: impaired consolidation of nonREM sleep at all circadian phases. Sleep. 2001;24(5):565–77.
97. Musiek ES, Holtzman DM. Mechanisms linking circadian clocks, sleep, and neurodegeneration. Science. 2016;354(6315):1004–8.
98. Videnovic A, Lazar AS, Barker RA, Overeem S. 'The clocks that time us'—circadian rhythms in neurodegenerative disorders. Nat Rev Neurol. 2014;10(12):683–93.
99. Cho JH, Bhutani S, Kim CH, Irwin MR. Anti-inflammatory effects of melatonin: A systematic review and meta-analysis of clinical trials. Brain Behav Immun. 2021;93:245–53.
100. McEwen BS, Karatsoreos IN. Sleep deprivation and circadian disruption: stress, allostasis, and allostatic load. Sleep Med Clin. 2015;10(1):1–10.
101. Dzierzewski JM, Donovan EK, Kay DB, Sannes TS, Bradbrook KE. Sleep inconsistency and markers of inflammation. Front Neurol. 2020;11:1042.
102. Ray S, Lach R, Heesom KJ, Valekunja UK, Encheva V, Snijders AP, et al. Phenotypic proteomic profiling identifies a landscape of targets for circadian clock-modulating compounds. Life Sci Alliance. 2019;2(6):e201900603.
103. Ribeiro RFN, Cavadas C, Silva MMC. Small-molecule modulators of the circadian clock: Pharmacological potentials in circadian-related diseases. Drug Discov Today. 2021;26(7):1620–41.
104. Aïd S, Bosetti F. Targeting cyclooxygenases-1 and -2 in neuroinflammation: Therapeutic implications. Biochimie. 2011;93(1):46–51.
105. Cao Y, Song Y, Ning P, Zhang L, Wu S, Quan J, et al. Association between tumor necrosis factor alpha and obstructive sleep apnea in adults: a meta-analysis update. BMC Pulm Med. 2020;20(1):215.
106. Ricciotti E, FitzGerald GA. Prostaglandins and inflammation. Arterioscler Thromb Vasc Biol. 2011;31(5):986–1000.
107. Sellix MT, Yoshikawa T, Menaker M. A circadian egg timer gates ovulation. Curr Biol. 2010;20(6):R266–7.
108. Zielinski MR, Kim Y, Karpova SA, McCarley RW, Strecker RE, Gerashchenko D. Chronic sleep restriction elevates brain interleukin-1 beta and tumor necrosis factor-alpha and attenuates brain-derived neurotrophic factor expression. Neurosci Lett. 2014;580:27–31.
109. Zielinski MR, Krueger JM. Sleep and innate immunity. Front Biosci (Schol Ed). 2011;3(2):632–42.
110. Brandt JA, Churchill L, Rehman A, Ellis G, Mémet S, Israël A, et al. Sleep deprivation increases the activation of nuclear factor kappa B in lateral hypothalamic cells. Brain Res. 2004;1004(1–2):91–7.
111. Chen Z, Gardi J, Kushikata T, Fang J, Krueger JM. Nuclear factor-kappaB-like activity increases in murine cerebral cortex after sleep deprivation. Am J Phys. 1999;276(6):R1812–8.
112. Liu T, Zhang L, Joo D, Sun S-C. NF-κB signaling in inflammation. Signal Transduct Target Ther. 2017;2:17023.
113. Pascual G, Glass CK. Nuclear receptors versus inflammation: mechanisms of transrepression. Trends Endocrinol Metab. 2006;17(8):321–7.
114. Lehnardt S, Schott E, Trimbuch T, Laubisch D, Krueger C, Wulczyn G, et al. A vicious cycle involving release of heat shock protein 60 from injured cells and activation of toll-like receptor 4 mediates neurodegeneration in the CNS. J Neurosci. 2008;28(10):2320–31.

115. Rivest S. Molecular insights on the cerebral innate immune system. Brain Behav Immun. 2003;17(1):13–9.
116. Panossian L, Fenik P, Zhu Y, Zhan G, McBurney MW, Veasey S. SIRT1 regulation of wakefulness and senescence-like phenotype in wake neurons. J Neurosci [Internet]. 2011;31(11) [cited 2023 Sept 15]. Available from: https://pubmed.ncbi.nlm.nih.gov/21411645/
117. Gomes Galvani M, Moreira Antunes HK, Monico-Neto M, Yujra VQ, Prado CM, Quintana HT, et al. Sleep deprivation interferes with JAK/STAT signaling pathway and myogenesis in the masseter muscle of rats. Med Princ Pract. 2021;30(3):253–61.
118. Lee JY, Jung E, Yeo H, Ahn SS, Lim Y, Lee YH. The natural janus kinase inhibitor agerarin downregulates interleukin-4-induced PER2 expression in HaCaT keratinocytes. Molecules. 2022;27(13):4205.
119. Havekes R, Abel T. The tired hippocampus: the molecular impact of sleep deprivation on hippocampal function. Curr Opin Neurobiol. 2017;44:13–9.
120. Wang X, Wang Z, Cao J, Dong Y, Chen Y. Melatonin alleviates acute sleep deprivation-induced memory loss in mice by suppressing hippocampal ferroptosis. Front Pharmacol. 2021;12:708645.
121. Berkowitz DE, White R, Li D, Minhas KM, Cernetich A, Kim S, et al. Arginase reciprocally regulates nitric oxide synthase activity and contributes to endothelial dysfunction in aging blood vessels. Circulation. 2003;108(16):2000–6.
122. Finger CE, Moreno-Gonzalez I, Gutierrez A, Moruno-Manchon JF, McCullough LD. Age-related immune alterations and cerebrovascular inflammation. Mol Psychiatry. 2022;27(2):803–18.
123. Mishra A, Bandopadhyay R, Singh PK, Mishra PS, Sharma N, Khurana N. Neuroinflammation in neurological disorders: pharmacotherapeutic targets from bench to bedside. Metab Brain Dis. 2021;36(7):1591–626.
124. Olivieri F, Rippo MR, Prattichizzo F, Babini L, Graciotti L, Recchioni R, et al. Toll like receptor signaling in "inflammaging": microRNA as new players. Immun Ageing. 2013;10(1):11.
125. Ungvari Z, Bailey-Downs L, Gautam T, Sosnowska D, Wang M, Monticone RE, et al. Age-associated vascular oxidative stress, Nrf2 dysfunction, and NF-{kappa}B activation in the nonhuman primate Macaca mulatta. J Gerontol A Biol Sci Med Sci. 2011;66(8):866–75.
126. Yuan X-Z, Sun S, Tan C-C, Yu J-T, Tan L. The role of ADAM10 in Alzheimer's disease. J Alzheimers Dis. 2017;58(2):303–22.
127. Wong SM, Jansen JFA, Zhang CE, Hoff EI, Staals J, van Oostenbrugge RJ, et al. Blood-brain barrier impairment and hypoperfusion are linked in cerebral small vessel disease. Neurology. 2019;92(15):e1669–77.
128. Dendrou CA, McVean G, Fugger L. Neuroinflammation—using big data to inform clinical practice. Nat Rev Neurol. 2016;12(12):685–98.
129. Di Benedetto S, Müller L, Wenger E, Düzel S, Pawelec G. Contribution of neuroinflammation and immunity to brain aging and the mitigating effects of physical and cognitive interventions. Neurosci Biobehav Rev. 2017;75:114–28.
130. Singh D. Astrocytic and microglial cells as the modulators of neuroinflammation in Alzheimer's disease. J Neuroinflamm. 2022;19(1):206.
131. Garden GA. Epigenetics and the modulation of neuroinflammation. Neurotherapeutics. 2013;10(4):782–8.
132. Garbarino S, Lanteri P, Bragazzi NL, Magnavita N, Scoditti E. Role of sleep deprivation in immune-related disease risk and outcomes. Commun Biol. 2021;4(1):1304.
133. Neves J, Sousa-Victor P. Regulation of inflammation as an anti-aging intervention. FEBS J. 2020;287(1):43–52.
134. Yun-Hee, Youm Thirumala-Devi, Kanneganti Bolormaa, Vandanmagsar Xuewei, Zhu Anthony, Ravussin Ayinuer, Adijiang John S., Owen Michael J., Thomas Joseph, Francis John S., Parks Vishwa Deep, Dixit (2012) The NLRP3 Inflammasome Promotes Age-Related Thymic Demise and Immunosenescence Cell Reports 1(1) 56-68 10.1016/j.celrep.2011.11.005

Sleep, Aging, and Inflammation: Unraveling the Connections

Yogendra Singh, Upendra Baitha, and Lal Chandra Vishwakarma

Abstract This chapter delves into the complex interrelationships among sleep, aging, and inflammation, illuminating the physiological mechanisms that link these core facets of human health. People's sleep patterns become more noticeable as they get older, with modifications to their circadian rhythms and sleep architecture. The reciprocal relationship between inflammation and sleep is given a lot of attention. The common chronic inflammation associated with aging is examined in conjunction with how sleep affects immune responses, possible consequences for age-related health conditions, and how disturbed sleep can lead to altered inflammatory cytokines. In addition to providing insights into the potential significance of prioritizing sleep for optimal well-being in the aging population, it is essential to present a thorough overview of the relationships between sleep, aging, and inflammation.

Keywords Sleep · Immune system · Aging · Biomarkers · Cell interaction Inflammation

1 Introduction

Sleep is a fundamental biological need and physiological process that is important for the regulation of emotion, optimally functioning cognition, and maintaining homeostasis across several physiological functions [1]. It may be defined as a

Y. Singh (✉) · U. Baitha
Department of Medicine, All India Institute of Medical Sciences, New Delhi, India

L. C. Vishwakarma
Division of Sleep Medicine, Department of Medicine, Perelman School of Medicine, University of Pennsylvania, Philadelphia, PA, USA

P. K. Jha, H. Bouaouda (eds.), *Circadian Rhythms, Sleep and Inflammation*, Progress in Inflammation Research 93,
https://doi.org/10.1007/978-3-031-74744-1_9

reversible state of unconsciousness manifested as sluggish eye movements, a specific sleeping posture, a heightened arousal threshold, a decreased sensitivity to outside stimuli, an extended reaction time, and diminished motor activity. This chapter will focus on investigating the intricate connections between sleep, inflammation, and aging. Further, the effects of these interconnections on overall health and wellness shall be discussed.

2 Sleep: A Basic Biological Process

Sleep is characterized as a reversible state of apparent unconsciousness that is behaviorally characterized by the closure of eyes and assumption of recumbent position during the process of falling asleep and culminates in reduced receptiveness to environmental stimuli at its onset [2]. Adequate sleep is crucial for the proper functioning of every tissue and system in the body, including the immune system, emotional regulation, cognitive function, and removing brain waste [3, 4]. Two different processes, known as homeostatic process and circadian process regulate the state of sleep and vigilance. Our brain's 24-h internal clock, known as the circadian clock, controls our cycles of wakefulness and drowsiness in response to changes in light levels and a number of other zeitgebers in our surroundings. The Earth's revolution on its axis shapes our physiology and behavior. The biological circadian rhythm was developed to assist people in adjusting to environmental changes and anticipating shifts in temperature, radiation, and food supply. Humans would not be able to optimize energy expenditure and internal physiology without this endogenous circadian clock. A key conceptual framework in sleep research has been the two-process model of sleep regulation, as mentioned above. [5]. Sleep seems to help the optimization of metabolic processes in accordance with the at the proper stage of the 24-h cycle in addition to its short-term, use-dependent role in enforcing rest and fasting.

3 Homeostatic process and System Producing Sleep

A subset of sleep-active neurons has been identified in the rat hypothalamus. These neurons are located in the anterior hypothalamus, basal forebrain, and ventrolateral preoptic nucleus (VLPO), also known as the intermediate nucleus of the preoptic area. It is believed that these neurons inhibit the major ascending monoaminergic arousal systems during sleep and actively promote sleep. Gamma-aminobutyric acid (GABA) neurotransmitters, which are inhibitory, are primarily responsible for inhibiting the wake-generating areas of the ascending reticular activation system

Table 1 Techniques for quantifying sleep

	Established methods		New methods
Sleep	Subjective methods	Questionnaires and sleep diaries	In-ear EEG
		Direct observation	
	Objective methods	Actigraphy & polysomnography (PSG)	
		Scalp EEG	
Sleepiness	Subjective methods	Questionnaires & Epworth Sleepiness Scale (ESS)	Visual Analogue Scale of Sleepiness (VAS)
	Objective methods	Multiple sleep latency test (MSLT)	
		Maintenance of wakefulness test (MWT)	
		Oxford sleep resistance test (OSLER)	

(ARAS) in neurons located in the brainstem projection. Only the beginning of sleep and the Non-Rapid Eye Movement (NREM) phase are affected by this activation. Another part of the hypothalamus that is active during NREM sleep and sleep beginning is the median preoptic area (MnPO) [6]. Sleep is therefore brought on by GABA neuronal activity and the suppression of alertness neurons [7]. The brain region between the caudal medulla and the rostral midbrain was also shown to produce Rapid Eye Movement (REM) sleep [8]. Animal studies have identified the neuronal system responsible for producing REM sleep, in which Electroencephalogram (EEG) signals were accompanied by muscular atonia [9] (Table 1).

4 The Stages of Sleep

Each stage of sleep has a distinct pattern of brain activity that can be measured using EEG. The NREM and REM stages are the two main sleep phases. For observed sleep depths, NREM sleep in humans is further divided into three stages, N1, N2, and N3 [10]. Dreaming occurs primarily during REM sleep, which is comparable to sleep in the awake brain. An eight-hour sleep cycle typically consists of four to six cycles, with each cycle lasting roughly 90–120 min. The first sleep cycle includes the entire duration from the patient's first beginning of sleep until the end of the first REM phase. Stage N3 sleep is more common in the early half of the night, and REM sleep becomes more common as the night wears on. The different stages of sleep are indicated by specific brain oscillations that are picked up by polysomnography. Slow oscillations of 0.5–4 Hz, slow spindles of 9–12 Hz, and fast spindles of 12–15 Hz are characteristics of

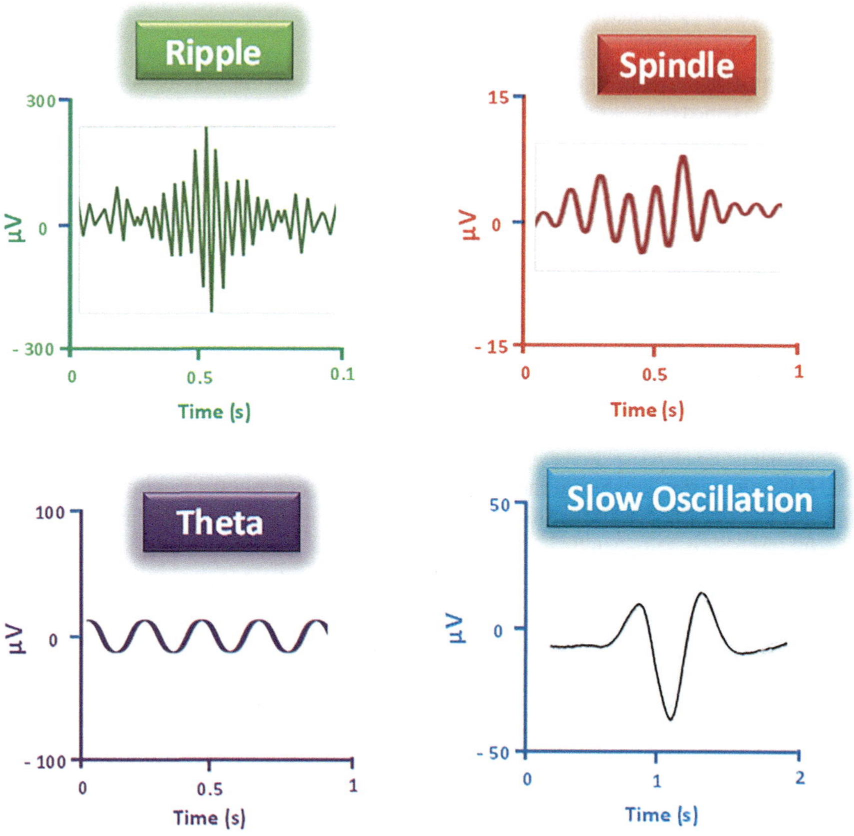

Fig. 1 Sleep oscillations and architecture [11, 12]

slow-wave sleep (SWS) [11, 12]. Theta oscillations are noticeable during NREM sleep (Fig. 1).

Stage N1 (5%): Theta waves, the lowest voltage, lightest sleep type, occur shortly after you fall asleep. *Stage N2 (45%)*: K complexes, slow-wave delta brain activity, sleep spindles, and deep sleep. *Stage N3 (25%)*: Greatest amplitude, delta brain activity, and deepest NREM sleep. *REM Sleep (25%)*: Eyelids and eyes may flutter, respiration may become uneven, and dream sleep may occur (Table 2).

Table 2 Adults sleep metrics and their attributes

Sleep metric	Definition	Normal characteristics
Sleep architecture	The basic structural organization of normal sleep. There are two types: NREM and REM sleep	Single cycles are approximately 90–120 min and four to six cycles occur during a typical 8-h night of sleep
Sleep efficiency index (SEI)	The ratio of the amount of time asleep to the amount of time spent in bed is expressed as a percentage	≥85–90%
Sleep onset latency (SOL)	The amount of time taken to fall asleep	≤20 min
Total sleep time (TST)	The amount of time spent asleep, either during the night or 24 h	7–9 h
Wake after sleep onset (WASO)	The amount of time spent awake during a sleep period after first falling asleep	Time awake 20–40 min during the night
Nocturnal awakenings	Frequency of awakenings per night	Few, less than 5% of TST
Arousals	Any abrupt change from deep sleep to a mixture of very light sleep and/or partial wakefulness	10–22 arousals per hour
Number/patterns of sleep cycles	Alternation of all sleep stages forming a sleep cycle	Five to six cycles per night, both organized and in sequence
Respiratory effort-related arousal (RERA)	A sequence of breaths lasting at least 10 s characterized by increased respiratory effort leading to an arousal from sleep	Young and middle-aged adults average around 15 arousals per hour

Functions of sleep [13, 14]

- Immune system regulation
- Reduction in energy use
- Restoration of brain energy stores
- Metabolite clearance from the brain
- Modulation of neuronal plasticity
- Restoration of cognitive performance

5 Sleep and Frailty

Frailty is a marker of aging and it is a common term found in geriatric literature [15]. Several definitions have been proposed which encompass a few major factors—impairment of function, energy and endurance, disability, and limitation of social roles [15]. It has a bidirectional relationship with the risk of multi-morbidity, which is defined as having two or more diseases [16]. Frailty is a marker of aging and sleep is associated with frailty, although the research in this area is scanty.

6 Sleep Duration and Sleep Quality

A systematic review including six studies showed that perceived sleep quality was associated with frailty [17, 18]. However, the sleep–wake pattern did not affect frailty in this systematic review [17] Later, meta-analyses showed that short sleep duration (<6 h/day) as well as long sleep duration (>8 h/day) was associated with frailty in older adults, irrespective of gender [19–21]. Insomnia symptoms did not influence frailty in a systematic review [17]. Another meta-analysis reported that any kind of insomnia—initial, middle, or terminal increased the risk of frailty among the elderly [18].

7 Sleep and Sarcopenia

Short (<6 h/day) as well as long sleep duration (>8 h/day) increases the risk of sarcopenia compared to those sleeping 6–8 h/day, especially among older adults, as well as those having poor sleep quality [20–23]. While the short sleep duration increases risk among both sexes, long sleep duration increases risk only in men and among all age groups [20–22]. Short (8h/day) increases the risk of sarcopenia compared to those sleeping 6-8 h/day, especially among older adults, as well as those having poor sleep quality. In contrast, the short sleep duration increases risk among both sexes, and the long sleep duration increases risk only in men and all age groups, people with short sleep duration are considered to be sleep deprived which in turn impairs rejuvenation and repair, increases insulin resistance, impairs protein synthesis, lead to chronic low-grade inflammation, promotes lack of physical activity, and increased risk of nicotine and alcohol use [22]. Similarly, long sleep duration is thought to increase the risk of sarcopenia through sleep fragmentation, lowering of physical activity, obesity, and chronic inflammation [22]. Furthermore, sarcopenia has been found to increase the risk of mortality [24]. Some studies have shown that obstructive sleep apnea increases the risk of sarcopenia; however, this is still preliminary data [25]. Apnea–hypopnea index has been positively correlated with skeletal muscle mass index and negatively with skeletal muscle density in another study [26].

8 Sleep and Mortality

In younger adults, long sleep duration was associated with greater inflammation even after controlling for age, gender, smoking, alcohol use, medical comorbidities, and medications [27, 28]. A meta-analysis reported that sleep disturbance and long sleep duration were associated with a higher burden of inflammatory markers (IL-6 and CRP but not TNF-alpha) while short sleep duration, experimental sleep deprivation, and sleep restriction did not increase inflammatory burden [29].

Hall et al. [30] reported that both short (<6 h/day) and long sleep (>8 h/day) durations increased the risk of mortality in older adults. On the other hand, Osteoporotic Fractures in men's sleep study showed that short sleep duration (<5 h/day), but not long sleep duration (>8 h/day) increased the risk of mortality among older men [31]. However, the magnitude of risk associated with sleep duration was attenuated when chronic health conditions and inflammatory markers (IL-6, TNF-alpha, and CRP) were taken into consideration suggesting that these factors were more important compared to sleep duration [30, 31]. However, these are individual studies which usually study a small population, and hence, results may not be generalizable.

A meta-analysis reported that irrespective of the definition of short sleep (4–7 h/day) and long sleep (>8–10 h/day) across studies, both increased the risk of mortality among all age groups, though the effect of long sleep in elderly was stronger [32]. Similar results were found in other meta-analyses that have included the studies conducted among the elderly population [33–35] who reported that short sleep (<7 h) did not increase risk of mortality while long sleep (>8 h) was associated with progressively increased risk for each hour.

The effect of sleep duration also differs between genders—while long sleep duration increases the risk for all-cause mortality in both genders, short sleep duration increases all-cause mortality only in men [35]. The association between sleep duration and mortality could also be the function of age and could run on a gradient of a stronger association among the elderly compared to the younger population [32, 33, 35]. Short sleep duration (<5 h/day) and long sleep duration (>8 h/day) among young adults increase the risk of mortality, but not in elderly aged more than 65 years [36].

Mortality has been assessed in the studies by dividing it into various subgroups, attributed to cardiovascular reasons, cancers, and by lumping all causes of mortality together (Table 3). Short sleep has been found to increase the risk of diabetes, obesity, hypertension, and coronary artery disease which could in turn lead to mortality and a linear association has been found between risk of all-cause mortality and sleep duration less than 6 h/day [37]. In young adults (<65 years), short sleep increased the risk of all-cause mortality as well as mortality due to cardiovascular disease, though a significant heterogeneity was observed in the studies assessing the effect of long sleep [38].

Thus, there are conflicting results across studies that could be influenced by variables related to the included population—demographics, medical history, definition of outcome variables, and, methodological differences in assessment of study variables, to name a few.

Table 3 Results of the meta-analysis which assessed the relationship between sleep duration and mortality

Author	Short sleep duration		Long sleep duration		Remarks
	All-cause mortality	Cause-specific mortality	All-cause mortality	Cause-specific mortality	
Gallicchio and Kalesan [39]	Increased risk	No effect on cardiovascular and cancer-related mortality	Increased risk	Increased risk of cardiovascular and cancer-related mortality	• Short sleep: <6–8 h • Long sleep: >8–10 h
Cappuccio et al. [32]	Increased risk	–	Increased risk	–	• Short sleep: <5–7 h • Long sleep: >8–9 h
da Silva et al. [33]	Increased risk	–	Increased risk	Increased risk of cardiovascular mortality	• Short sleep: <5–7 h • Long sleep: 8– >11 h
Kwok et al. [34]	No effect	–	Increased risk	–	• Short sleep: <7 h • Long sleep: >8 h
García-Perdomo et al. [40]	No association found	–	Increased risk	–	• Short sleep: <7 h • Long sleep: >9 h
Pienaar et al. [38]	Increased risk	Increased risk of cardiovascular mortality	Increased risk	–	• Short sleep: <6 h • Long sleep: ≥8 h

9 Effect of Sleep Timing

Not only the duration of sleep but also the timing of the sleep is important for health. Persons with evening chronotype and morning chronotype have similar sleep architecture when they are allowed to sleep according to their preferred sleep–wake schedule [41]. However, maintaining a sleep–wake schedule at non-natural timing can alter the proportion of different sleep stages that one gets during the naturally timed sleep, the preponderance of slow wave sleep in the early night, and the dominance of REM sleep toward morning [42]. Shift workers can be divided into two major types—those having rotating shifts and those working permanently in either day or night shifts. Those working rotating shifts and those working in day shifts spend at least some time sleeping according to their natural sleep–wake schedule compared to those working in night shift permanently. Permanent night work, but not shift work, has been found to increase the risk of all-cause mortality and death due to cardiovascular diseases but not cancer-related mortality [43].

10 Aging: Effects on Sleep and Protein Homeostatic Regulation

The quality of sleep varies with a healthy aging process. Studies have shown that as people get older, they tend to require less sleep [44]. Some of these alterations include decreased melatonin levels, shortened sleep duration, reduced deep slow-wave sleep, increased sleepiness during the day, worse sleep consolidation by greater awakening, and increased sleep onset latency [45]. Notably, age-related changes in sleep features are observed in several studies [44, 46]. The idea that sleep has been retained throughout evolution because it is necessary is supported by the fact that many species need sleep even though this requirement decreases with age. Age-related variations in sleep characteristics could potentially be a contributing cause of cognitive deterioration. Some studies have linked insufficient slow-wave oscillations and sleep spindles during NREM sleep to memory loss and forgetfulness in the elderly [47]. Research has revealed that middle-aged people's sleep quality may impair their cognitive function in later life. Therefore, it seems that executive function and memory functions are where sleep quality affects midlife cognition. Improved midlife sleep quality may help to mitigate subtle deficiencies in these age-sensitive cognitive skills, which may suggest an increased risk for cognitive decline later in life [48].

The population's sleep duration has decreased recently. Furthermore, phenotypic age and sleep duration had an inverse U-shaped connection. Furthermore, longer sleep duration is linked to the advantages of consistent, regular exercise, whereas shorter sleep duration and more frequent exercise are linked to phenotypic aging [49]. Short and long sleep durations were positively connected with abnormal CRP, HDL, HbA1c, blood glucose, and insulin levels, whereas sleep disturbance was linked to aberrant triglyceride and blood glucose levels. Since sleep is a controllable component, maintaining optimal sleeping habits can help maintain a balanced metabolism and reduce the risk of developing cardiovascular disease [50] (Table 4).

Table 4 Potential aging biomarkers identified in genomics, transcriptomics, proteomics, and metabolomics studies

Omics	Biomarkers	Function/application	References
Genomics	DNA methylation aging clocks	Biological age estimation method	[51]
	DNA methylation grim age	Been correlated with diseases and can predict mortality	[52]
	DNAm pattern of 353 CpG sites	Estimate physiological aging	[53]
	Forkhead box O3 gene (FOXO3)	Related to prolonged lifespan	[54–56]
Transcriptomics	Transcriptomics aging clocks	Predictors of age	[57]
	The number of ABCG1	Determines human lifespan	[58, 59]
	The expression of 11 genes (AMZ1, MANEAL, PARP3, KIAA0408, ISM1, CRIP1, NEFL, PHLDA3, DDB2, CHN1, CAPN2)	Positively correlated with aging	[60]
	miR-22-3p and miR-28-3p	Positively correlated with age	[61]
	miR-181a, miR-434-3p, miR-431, miR-29, and miR-126	In sarcopenia	[62]
	Telomere-lncRNA	Can regulate the telomerase activity and survival rate of neural stem cells during aging	[63]
Proteomics	Proteomics aging clocks	Accurately predict the age of a person	[64]
	GDF15, PTN, ADAMTS5, FSHB, SOST, CHRDL1, NPPB, EFEMP1, MMP12, and CTSV	Related to aging	[65]
	LGALS3BP, MASP2, DNASE1, ANPEP, IGFBP1	Assess the rate of aging	[66]
	CLEC3B, CRISP3, IGFAS, TAS1R3, and TGFBI	Be related to healthy aging	[67]
	Serine protease inhibitors, SCT1, and GDF15	As biomarkers of aging	[68]
	Methionine sulfoxide	A marker of biological aging	[69]
Metabolomics	CoA catabolism, vitamin E, lysine, tryptophan, and tyrosine metabolism	Related to aging	[70]
	Monoacylglycerides, diacylglycerols, and phosphoserine	Show a decreasing trend with age	[71]
	The product of proteolysis and L-γ-glutamyl-L-leucine	Increases independently of gender during aging	[71]

11 Sleep Throughout Life

As individuals age, humans and many other species undergo substantial changes in the kind, quantity, and temporal arrangement of their sleep [2, 72]. When children reach primary school age, their sleep patterns become monophasic from polyphasic throughout the early years. Once men and women reach their early 20s, the timing shifts reversely, drifting back to earlier times. It was shown that young people slept for 8.9 h during their terminal illness, compared to 7.5 h for older adults [73]. REM sleep accounts for only 20% of a child's entire sleep time by the time they are in elementary school, with a large increase in SWS [74]. 50% of the sleep a baby gets in the early stages of infancy is REM sleep. Puberty coincides with a decrease in cortical layer slow wave activity (SWA) and SWS. There is a 65% drop in SWA between the ages of 11 and 17 [75]. There is a persistent reduction in older persons; the projected decline between 45 and 90 years of age is approximately one-third of the decline observed between 20 and 40 years of age [76].

The Phillips–Robinson Model The model consists of a core flip-flop system comprising mutually inhibitory wake-activating (monoaminergic, MA) and sleep-active ventrolateral preoptic region neurons. The firing rate of wake active neurons (QMA) increases homeostatic sleep pressure (H), and the parameter m, which gauges the rate of homeostatic rise during wake, is a measure of sleep requirement. The ventrolateral preoptic area (VLPO) is impacted by a sleep drive with circadian (C) and homeostatic (H) processes (nvh and nvc strengths, respectively). Hence, nvc stands for the circadian amplitude, and nvh for the homeostatic process' sensitivity (Fig. 2). The circadian pacemaker's natural period (TC) is 24.2 h, but photoreceptors synchronize it with the outside lighted-dark cycle [77].

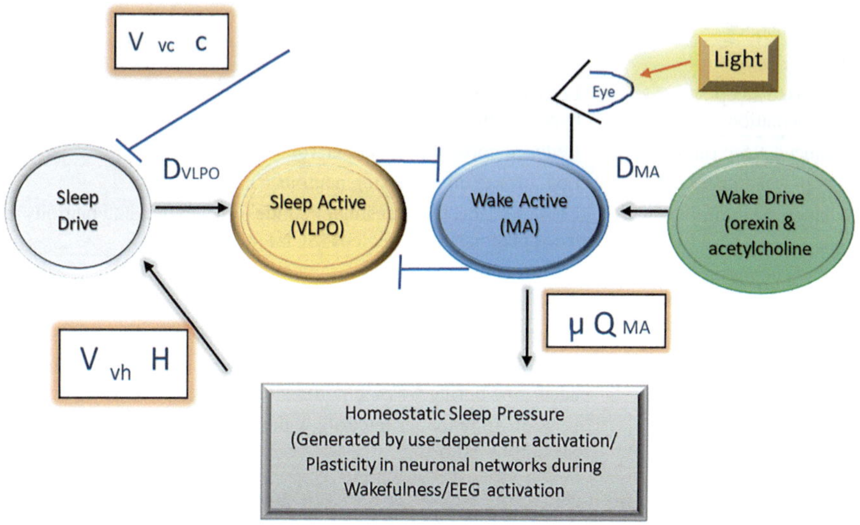

Fig. 2 Phillips–Robinson model [77]

12 Age-Related Sleep Changes

During the first few months of life, the sleep architecture of both male and female newborns alters in similar ways. Neonates and early infants have an immature and chaotic sleep structure compared to children because their underlying brain structures and sleep regulation systems are still growing. The minimal amount of sleep that is necessary for each age group typically decreases as we get older and increases for small children. While these patterns are not universal, most individuals can relate to the fact that infants sleep the most hours, and the elderly sleep the least. Aging was associated with decreased TST, more awakenings, and less productive sleep regardless of menopause and nocturnal melatonin release specifically decreases with menopause [78] (Tables 5 and 6).

Table 5 Recommended hours of sleep in different age groups by the National Sleep Foundation

Age group	Recommended sleep (h)	Insufficient sleep (h)	Oversleeping (h)
Newborns (0–3 months)	14–17	≤11	≥19
Infants (4–11 months)	12–15	≤10	≥18
Toddlers (1–2 years)	11–14	≤9	≥16
Preschoolers (3–5 years)	10–13	≤8	≥14
School children (6–13 years)	9–12	≤7	≥12
Teens (14–17 years)	8–10	≤7	≥11
Young adults (18–25 years)	7–9	≤6	≥11
Adults (26–64 years)	7–9	≤6	≥10
Seniors (65 years & older)	7–8	≤5	≥9

Table 6 The effect of normal aging on sleep [79]

Sleep changes	Potential neurobiological explanation
Macro change	
Increased sleep fragmentation	Loss of galanin expressed in the preoptic area of the hypothalamus
Reduced sleep duration	Reduced cortical gray matter volume
Sleep–wake instability	Loss of orexin-expressing neurons
Change sleep timing	Loss of Vasoactive Intestinal Peptide (VIP) expressing neurons in the suprachiasmatic nucleus (SCN)
Micro changes	
Impaired SWS oscillations	Grey matter volume loss in frontal regions
Reduced sleep spindles	Hippocampal volume loss
Reduced SWS rebound (prolonged wake)	Loss of adenosine receptors, reducing responsiveness to rising adenosine during wake, glial function changes
Impaired sleep spindle expression	Loss of white matter integrity within the body and splenium of the corpus callosum

13 The Sleep and Inflammation

13.1 Circadian Rhythm's Impact on Immune Function

Sleep and the circadian rhythm have a major impact on immune system activities. Research on the sleep–wake cycle shows that immune factors, like the production of proinflammatory cytokines and the counts of undifferentiated naïve T cells, peak in early nocturnal sleep, while circulating immune cells like anti-inflammatory cytokines and cytotoxic natural killer (NK) cells (Fig. 3) peak in activity during the day [80].

Adipocytes, neutrophils, monocytes, macrophages, and endothelial cells can all generate cytokines. It should be mentioned that the amount of each cytokine that contributes to circulating levels is uncertain, especially when sleep deprived [81]. Research demonstrating that decreasing nightly sleep to 4 h increased monocyte production of interleukin-6 (IL-6) and tumor necrosis factor-alpha (TNF-α) is necessary to determine the source of inflammation linked to sleep loss [82]. Monocytes are powerful makers of TNF-α and IL-6, but they are not the only ones; sleep deprivation activates a molecular pathway (Fig. 4) that causes the creation of these chemicals [83].

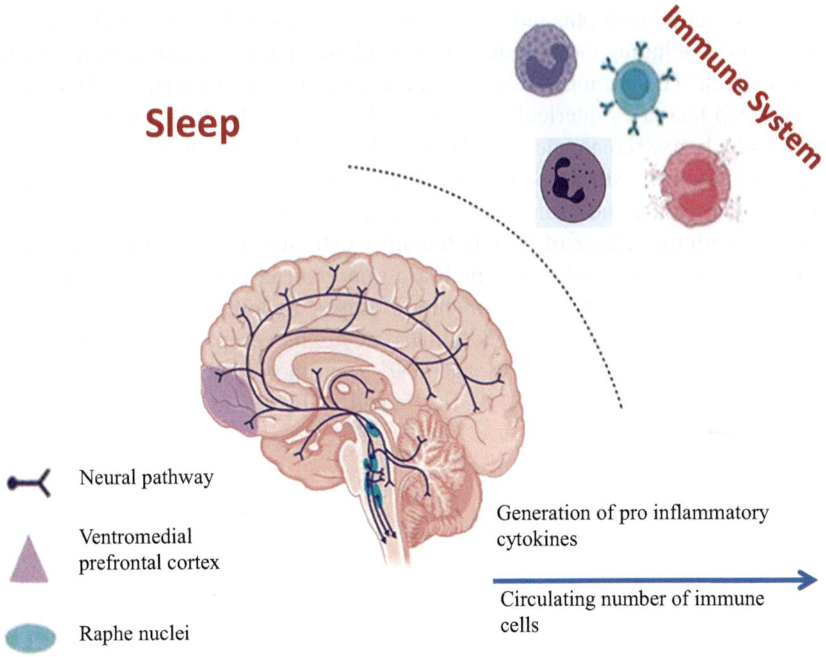

Fig. 3 Sleep and immune system [80]

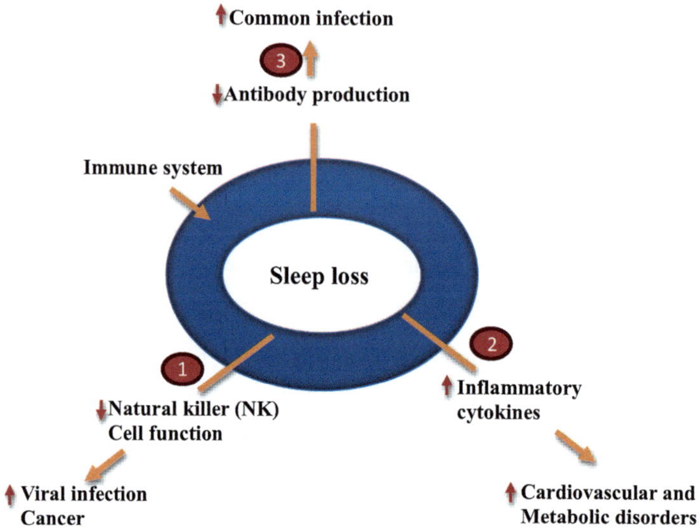

Fig. 4 Effects of sleep loss on the immune system [82]

14 Sleep and Immune System

The quality, duration, depth, and disruption of sleep can all be influenced by a variety of factors, including the circadian rhythm, physical activity, and the homeostatic desire to sleep. The immune system communicates mostly through cytokines, and lack of sleep increases interleukin levels of IL-6, IL-1β, IL-2, IL-18, and TNF-α, some of which may remain elevated for a week [84]. The immune system can cause sleep disruptions and sleep disruptions can contribute to a loss in the immune system's capacity to respond to an invasion. An increase in sleep disturbances has been associated with the release of proinflammatory cytokines (Fig. 5). The brain's glia and neurons generate cytokines, which are important in controlling the circadian rhythm in areas such as the brainstem, hippocampus, hypothalamus, and neocortex. In these areas, astrocytes and neurons both have cytokine receptors. Proinflammatory cytokines (IL-1β, IL-2, IL-6, IL-18, and TNF-α) promote NREM sleep, while anti-inflammatory cytokines (IL-4, IL-10, IL-13, and TGF) suppress it [85].

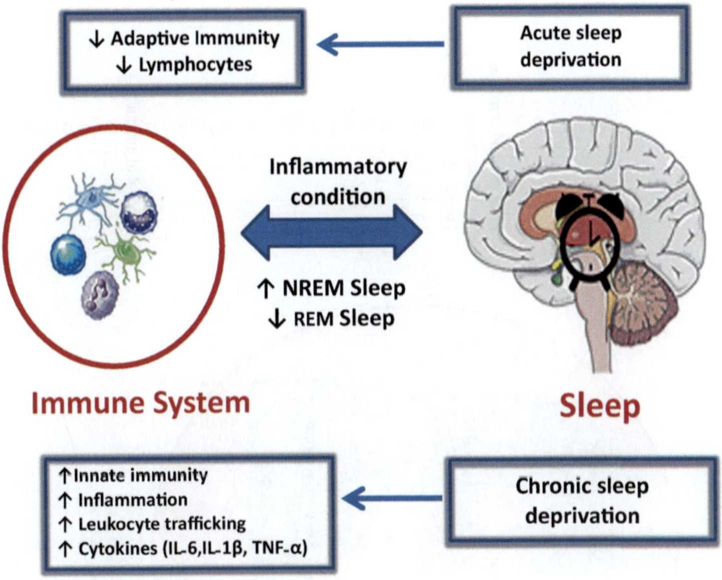

Fig. 5 Interaction between sleep and the immune system [85]

15 Sleep Deprivation and Inflammatory Markers

When IL-1 and TNF-α are administered intraventricularly to rats, NREM sleep increases by 60–70% [86]. When mice's sleep patterns return to normal, some of the inflammatory effects of sleep deprivation persist. It turns out that when sleep is disrupted, inflammation influences not only the quantity and quality of inflammatory mediators but also their levels and activities.

16 Inflammation and Disorders of Sleep

Multiple inflammatory indicators increase in response to sleep deprivation. However, mediators such as TNF-α, IL-6, and C-reactive protein (CRP) appear to be the most significant in human research [87, 88]. CRP levels were greater in young individuals with symptoms of sleeplessness [89]. Patients who reported sleeping for more than 10 h, on the other hand, had significantly higher levels of CRP and IL-6. This effect may contribute to the explanation of why patients with sleep disturbances are more likely to suffer from inflammatory diseases such as arterial hypertension or diabetes mellitus [27, 28].

Peripheral expression of antimicrobial peptides and inflammatory cytokines occurs in response to several forms of organismal stress, such as acute sleep deprivation, acute inflammatory exposure, and cellular stress (Fig. 6). Similarly, acute

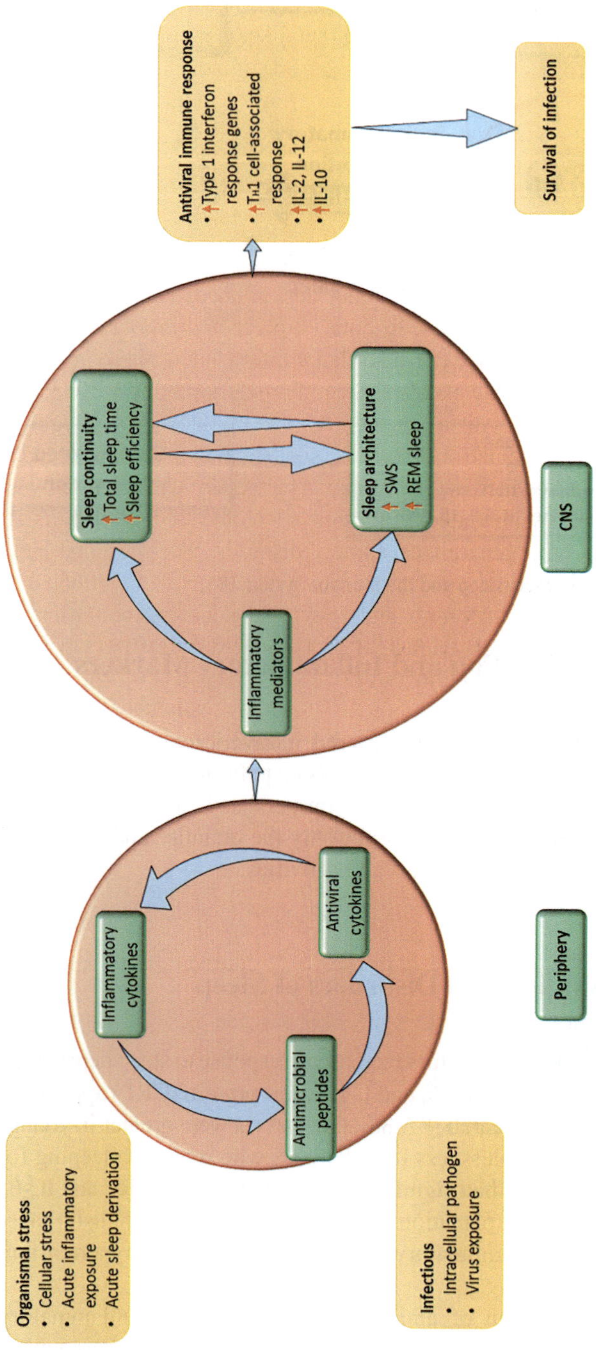

Fig. 6 Potential mechanisms affect the immune system's involvement in sleep

infectious exposure to an intracellular pathogen activates inflammatory or antiviral cytokine profiles. These peripheral signals are picked up by the central nervous system, which then utilizes them to cause the brain's inflammatory cytokine activity. This alters sleep patterns, extending the duration of sleep and its efficiency overall, as well as reducing the duration of REM sleep and lengthening the time spent in SWS. Increases in antiviral immune response profiles, which are associated with increases in sleep continuity, SWS, and T helper 1 cell, may have an effect on improved survival in the event of acute infection [90–93].

17 Sleep and Immune-Related Disease

Infection Longer sleep durations are associated with significantly reduced parasitism levels, and lower sleep quality is associated with higher infection risk and negative infection outcomes [27, 28, 94, 95]. In addition, teens who sleep for shorter durations of time roughly 6 h are more susceptible to catching common colds, the flu, gastroenteritis, and other infectious diseases than teenagers who sleep for longer durations roughly 7 h [96].

Neurodegenerative Diseases (NDD) Age-related disorders that target specific populations of neurons in the central nervous system are called NDDs; these include Alzheimer's disease, multiple sclerosis, Parkinson's disease, Huntington's disease, and amyotrophic lateral sclerosis. However, studies conducted on humans and animals have shown that altered circadian rhythm/clock genes may be a contributing factor to the development of NDDs [97–99].

Autoimmune Diseases (AD) Ankylosing spondylitis, rheumatoid arthritis, and systemic lupus erythematosus are among the autoimmune illnesses that have been associated with insomnia [100].

Vascular and Metabolic Disorders Sleep deprivation is associated with an increased risk of cardiovascular disease (CVD) outcomes, such as a 15% increased risk of stroke [101], a 12% increased risk of all-cause mortality, and a 48% increased risk of coronary heart disease [32, 102]. Additionally, it is linked to an increased subclinical atherosclerotic burden, which is the main underlying cause of cardiovascular diseases [103].

Cancer The shorter duration of nocturnal secretion of melatonin, which exerts anticancer properties through antimitotic, antioxidant, apoptotic, anti-estrogenic, and anti-angiogenic mechanisms, has led to the recognition of short sleep duration as a risk factor for several cancers, including colorectal, prostate, breast, and thyroid cancer [104–110].

However, there is not enough research evaluating cancer-related mortality, so it's impossible to draw firm conclusions. Although OSA has been shown to raise the risk of malignant tumors overall, there is not enough information to determine whether it also raises the risk of any particular malignancy [111]. The risk of incidental cancer is higher in older patients (over 60), but it is not influenced by gender. It is widely acknowledged that OSA tends to increase cardiovascular mortality whereas PAP therapy tends to decrease it; nevertheless, the extent of the effects varies widely and is not characterized in females or across age groups [112]. It has been discovered that sleeplessness considerably raises the risk of hypertension [113], metabolic syndrome [114], and cerebrovascular diseases [115]. There are complex connections between sleep duration and disorders linked to aging. In a middleaged and young-old adult population, the study found that persons who sleep more than 8 hours have shorter leukocyte telomere length (LTL) than those who sleep 7–8 hours. Those with an evening chronotype, men, and older adults are more likely to have a negative correlation between LTL and prolonged sleep duration [116].

18 Immune Systems and Sleep Deprivation

People who sleep less are more susceptible to long-term inflammatory disorders because sleep deprivation has been demonstrated to impair the immune system's capacity to combat inflammation through a variety of methods (Fig. 7). Regularly skipping short stretches of sleep has been related to altered antigen presentation, reduced Th1 response, increased Th2 response, and decreased production of

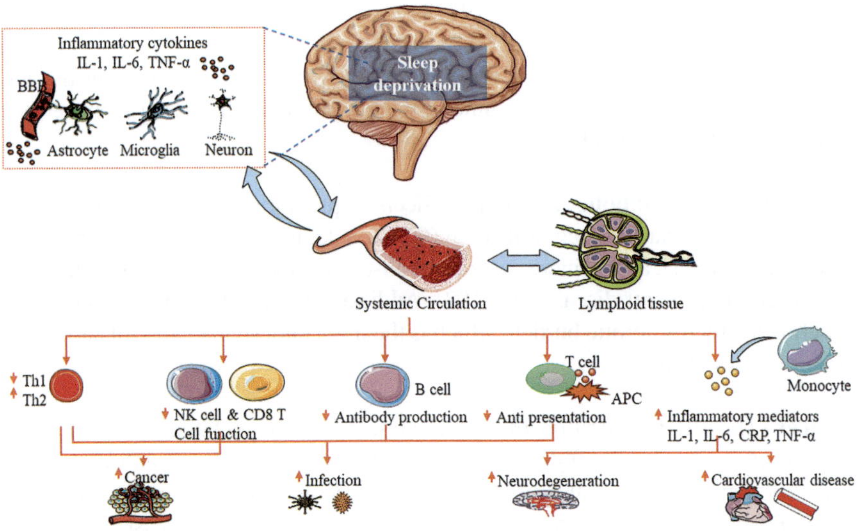

Fig. 7 Immunity-damaging effects of sleep loss

antibodies. It has also been linked to alterations in the circulating numbers and activity of particular cell subsets and total leukocytes, as well as elevated levels of proinflammatory markers and cytokines in the brain and other tissues.

Furthermore, the immune modulation associated with sleep deprivation may be influenced by altered monocyte response to immunological stimuli such as lipopolysaccharide. This finding lends credence to the theoretical associations between immune dysregulation resulting from sleep deprivation and the risk of immune-related disorders, such as those that are infectious, cardiovascular, metabolic, neurodegenerative, and neoplastic.

19 Conclusion

Chronic sleep deprivation in particular can cause an increase in inflammatory markers, which can negatively impact sleep. This could start a vicious cycle that quickens the aging process and the onset of age-related diseases. Maintaining healthy sleep habits, such as obtaining enough sleep for the right amount of time and quality, may help improve overall well-being and lessen aging-related inflammation. It is important to keep in mind that various sleep patterns and their effects on inflammation cause varied reactions in different persons. These connections may be impacted by variations in genetics, general health, and lifestyle. The relationship between sleep, aging, and inflammation is yet unknown. Nonetheless, there is data to bolster the theory that promoting restful sleeping habits could lessen inflammation associated with aging and enhance overall health. Further study is needed to develop targeted interventions to enhance sleep and reduce inflammation in the aging population, which would improve general public health. This will help us better understand the mechanisms behind these interactions.

Acknowledgments We thank Mahi Yadav and Dhanvi Yadav for their help searching articles and creating flow diagrams.

Compliance with Ethical Standards

Ethical approval This article does not contain any studies with human participants or animals performed by any authors.

Informed consent None.

Conflict of interest The authors declare that they have no conflict of interest.

Funding Source None.

References

1. Lane JM, Qian J, Mignot E, Redline S, Scheer FAJL, Saxena R. Genetics of circadian rhythms and sleep in human health and disease. Nat Rev Genet. 2023;24(1):4–20. https://doi.org/10.1038/s41576-022-00519-z. Epub 26 Aug 2022. PMID: 36028773; PMCID: PMC10947799
2. Siegel JM. Clues to the functions of mammalian sleep. Nature. 2005;437(7063):1264–71. https://doi.org/10.1038/nature04285. PMID: 16251951; PMCID: PMC8760626
3. Benveniste H, Lee H, Volkow ND. The glymphatic pathway: waste removal from the CNS via cerebrospinal fluid transport. Neuroscientist. 2017;23(5):454–65. https://doi.org/10.1177/1073858417691030. Epub 2 Feb 2017. PMID: 28466758; PMCID: PMC5547012
4. Zielinski MR, McKenna JT, McCarley RW. Functions and mechanisms of sleep. AIMS Neurosci. 2016;3(1):67–104. https://doi.org/10.3934/Neuroscience.2016.1.67. Epub 21 Apr 2016. PMID: 28413828; PMCID: PMC5390528
5. Borbély AA, Daan S, Wirz-Justice A, Deboer T. The two-process model of sleep regulation: a reappraisal. J Sleep Res. 2016;25(2):131–43. https://doi.org/10.1111/jsr.12371. Epub 14 Jan 2016. PMID: 26762182
6. Suntsova N, Szymusiak R, Alam MN, Guzman-Marin R, McGinty D. Sleep-waking discharge patterns of median preoptic nucleus neurons in rats. J Physiol. 2002;543(Pt 2):665–77. https://doi.org/10.1113/jphysiol.2002.023085. PMID: 12205198; PMCID: PMC2290500
7. Eriksson KS, Sergeeva O, Brown RE, Haas HL. Orexin/hypocretin excites the histaminergic neurons of the tuberomammillary nucleus. J Neurosci. 2001;21(23):9273–9. https://doi.org/10.1523/JNEUROSCI.21-23-09273.2001. PMID: 11717361; PMCID: PMC6763926
8. Luppi PH. Obituary: Michel Jouvet (1925-2017), The father of paradoxical sleep. J Sleep Res. 2017;26(6):832–4. https://doi.org/10.1111/jsr.12646. PMID: 29178163
9. Scammell TE, Arrigoni E, Lipton JO. Neural circuitry of wakefulness and sleep. Neuron. 2017;93(4):747–65. https://doi.org/10.1016/j.neuron.2017.01.014. PMID: 28231463; PMCID: PMC5325713
10. Berry RB, Brooks R, Gamaldo C, Harding SM, Lloyd RM, Quan SF, Troester MT, Vaughn BV. AASM scoring manual updates for 2017 (Version 2.4). J Clin Sleep Med. 2017;13(5):665–6. https://doi.org/10.5664/jcsm.6576. PMID: 28416048; PMCID: PMC5406946
11. Irwin MR. Why sleep is important for health: a psychoneuroimmunology perspective. Annu Rev Psychol. 2015;66:143–72. https://doi.org/10.1146/annurev-psych-010213-115205. Epub 21 July 2014. PMID: 25061767; PMCID: PMC4961463
12. Mölle M, Bergmann TO, Marshall L, Born J. Fast, and slow spindles during the sleep slow oscillation: disparate coalescence and engagement in-memory processing. Sleep. 2011;34(10):1411–21. https://doi.org/10.5665/SLEEP.1290. PMID: 21966073; PMCID: PMC3174843
13. Llorens F, Zarranz JJ, Fischer A, Zerr I, Ferrer I. Fatal familial insomnia: clinical aspects and molecular alterations. Curr Neurol Neurosci Rep. 2017;17(4):30. https://doi.org/10.1007/s11910-017-0743-0. PMID: 28324299
14. Krueger JM, Frank MG, Wisor JP, Roy S. Sleep function: Toward elucidating an enigma. Sleep Med Rev. 2016;28:46–54. https://doi.org/10.1016/j.smrv.2015.08.005. Epub 28 Aug 2015. PMID: 26447948; PMCID: PMC4769986
15. Sobhani A, Fadayevatan R, Sharifi F, Kamrani AA, Ejtahed HS, Hosseini RS, Mohamadi S, Fadayevatan A, Mortazavi S. The conceptual and practical definitions of frailty in older adults: a systematic review. J Diabetes Metab Disord. 2021;20(2):1975–2013. https://doi.org/10.1007/s40200-021-00897-x. PMID: 34900836; PMCID: PMC8630240
16. Vetrano DL, Palmer K, Marengoni A, Marzetti E, Lattanzio F, Roller-Wirnsberger R, Lopez Samaniego L, Rodríguez-Mañas L, Bernabei R, Onder G, Joint Action ADVANTAGE WP4

Group. Frailty and multimorbidity: a systematic review and meta-analysis. J Gerontol A Biol Sci Med Sci. 2019;74(5):659–66. https://doi.org/10.1093/gerona/gly110. PMID: 29726918

17. Wai JL, Yu DS. The relationship between sleep-wake disturbances and frailty among older adults: A systematic review. J Adv Nurs. 2020;76(1):96–108. https://doi.org/10.1111/jan.14231. Epub 21 Oct 2019. PMID: 31588595
18. Wen Q, Yan X, Ren Z, Wang B, Liu Y, Jin X. Association between insomnia and frailty in older population: A meta-analytic evaluation of the observational studies. Brain Behav. 2023;13(1):e2793. https://doi.org/10.1002/brb3.2793. Epub 14 Dec 2022. PMID: 36519213; PMCID: PMC9847606
19. Sun R, Xie Y, Jiang W, Wang E, Li X. Effects of different sleep disorders on frailty in the elderly: a systematic review and meta-analysis of observational studies. Sleep Breath. 2023;27(1):91–101. https://doi.org/10.1007/s11325-022-02610-5. Epub 13 Apr 2022. PMID: 35416613
20. Pourmotabbed A, Boozari B, Babaei A, Asbaghi O, Campbell MS, Mohammadi H, Hadi A, Moradi S. Sleep and frailty risk: a systematic review and meta-analysis. Sleep Breath. 2020;24(3):1187–97. https://doi.org/10.1007/s11325-020-02061-w. Epub 25 Mar 2020. PMID: 32215833
21. Pourmotabbed A, Ghaedi E, Babaei A, Mohammadi H, Khazaie H, Jalili C, Symonds ME, Moradi S, Miraghajani M. Sleep duration and sarcopenia risk: a systematic review and dose-response meta-analysis. Sleep Breath. 2020;24(4):1267–78. https://doi.org/10.1007/s11325-019-01965-6. Epub 12 Dec 2019. PMID: 31832982
22. Li X, He J, Sun Q. Sleep duration and sarcopenia: an updated systematic review and meta-analysis. J Am Med Dir Assoc. 2023;24(8):1193–206.e5. https://doi.org/10.1016/j.jamda.2023.04.032. Epub 7 June 2023. PMID: 37295459
23. Rubio-Arias JÁ, Rodríguez-Fernández R, Andreu L, Martínez-Aranda LM, Martínez-Rodriguez A, Ramos-Campo DJ. Effect of sleep quality on the prevalence of sarcopenia in older adults: a systematic review with meta-analysis. J Clin Med. 2019;8(12):2156. https://doi.org/10.3390/jcm8122156. PMID: 31817603; PMCID: PMC6947616
24. Xu J, Wan CS, Ktoris K, Reijnierse EM, Maier AB. Sarcopenia is associated with mortality in adults: a systematic review and meta-analysis. Gerontology. 2022;68(4):361–76. https://doi.org/10.1159/000517099. Epub 27 July 2021. PMID: 34315158
25. Tao X, Niu R, Lu W, Zeng X, Sun X, Liu C. Obstructive sleep apnea (OSA) is associated with increased risk of early-onset sarcopenia and sarcopenic obesity: Results from NHANES 2015-2018. Int J Obes (Lond). 2024;48(6):891–9. https://doi.org/10.1038/s41366-024-01493-8. Epub 21 Feb 2024. PMID: 38383717
26. Matsumoto T, Tanizawa K, Tachikawa R, Murase K, Minami T, Inouchi M, Handa T, Oga T, Hirai T, Chin K. Associations of obstructive sleep apnea with truncal skeletal muscle mass and density. Sci Rep. 2018;8(1):6550. https://doi.org/10.1038/s41598-018-24750-z. PMID: 29695811; PMCID: PMC5916913
27. Prather AA, Vogelzangs N, Penninx BW. Sleep duration, insomnia, and markers of systemic inflammation: results from the Netherlands Study of Depression and Anxiety (NESDA). J Psychiatr Res. 2015;60:95–102. https://doi.org/10.1016/j.jpsychires.2014.09.018. Epub 26 Sept 2014. PMID: 25307717; PMCID: PMC4250403
28. Prather AA, Janicki-Deverts D, Hall MH, Cohen S. Behaviorally assessed sleep and susceptibility to the common cold. Sleep. 2015;38(9):1353–9. https://doi.org/10.5665/sleep.4968. PMID: 26118561; PMCID: PMC4531403
29. Irwin MR, Olmstead R, Carroll JE. Sleep disturbance, sleep duration, and inflammation: a systematic review and meta-analysis of cohort studies and experimental sleep deprivation. Biol Psychiatry. 2016;80(1):40–52. https://doi.org/10.1016/j.biopsych.2015.05.014. Epub 1 June 2015. PMID: 26140821; PMCID: PMC4666828
30. Hall MH, Smagula SF, Boudreau RM, Ayonayon HN, Goldman SE, Harris TB, Naydeck BL, Rubin SM, Samuelsson L, Satterfield S, Stone KL, Visser M, Newman AB. Association between sleep duration and mortality is mediated by markers of inflammation and health

in older adults: the health, aging and body composition study. Sleep. 2015;38(2):189–95. https://doi.org/10.5665/sleep.4394. PMID: 25348127; PMCID: PMC4288599

31. Smagula SF, Stone KL, Redline S, Ancoli-Israel S, Barrett-Connor E, Lane NE, Orwoll ES, Cauley JA, Osteoporotic Fractures in Men (MrOS) Research Group. Actigraphy- and polysomnography-measured sleep disturbances, inflammation, and mortality among older men. Psychosom Med. 2016;78(6):686–96. https://doi.org/10.1097/PSY.0000000000000312. PMID: 26894325; PMCID: PMC4765903
32. Cappuccio FP, D'Elia L, Strazzullo P, Miller MA. Sleep duration and all-cause mortality: a systematic review and meta-analysis of prospective studies. Sleep. 2010;33(5):585–92. https://doi.org/10.1093/sleep/33.5.585. PMID: 20469800; PMCID: PMC2864873
33. da Silva AA, de Mello RG, Schaan CW, Fuchs FD, Redline S, Fuchs SC. Sleep duration and mortality in the elderly: a systematic review with meta-analysis. BMJ Open. 2016;6(2):e008119. https://doi.org/10.1136/bmjopen-2015-008119. PMID: 26888725; PMCID: PMC4762152
34. Kwok CS, Kontopantelis E, Kuligowski G, Gray M, Muhyaldeen A, Gale CP, Peat GM, Cleator J, Chew-Graham C, Loke YK, Mamas MA. Self-reported sleep duration and quality and cardiovascular disease and mortality: a dose-response meta-analysis. J Am Heart Assoc. 2018;7(15):e008552. https://doi.org/10.1161/JAHA.118.008552. PMID: 30371228; PMCID: PMC6201443
35. He M, Deng X, Zhu Y, Huan L, Niu W. The relationship between sleep duration and all-cause mortality in older people: an updated and dose-response meta-analysis. BMC Public Health. 2020;20(1):1179. https://doi.org/10.1186/s12889-020-09275-3. PMID: 32723316; PMCID: PMC7389345
36. Åkerstedt T, Ghilotti F, Grotta A, Bellavia A, Lagerros YT, Bellocco R. Sleep duration, mortality and the influence of age. Eur J Epidemiol. 2017;32(10):881–91. https://doi.org/10.1007/s10654-017-0297-0. Epub 30 Aug 2017. PMID: 28856478; PMCID: PMC5680380
37. Itani O, Jike M, Watanabe N, Kaneita Y. Short sleep duration and health outcomes: a systematic review, meta-analysis, and meta-regression. Sleep Med. 2017;32:246–56. https://doi.org/10.1016/j.sleep.2016.08.006. Epub 26 Aug 2016. PMID: 27743803
38. Pienaar PR, Kolbe-Alexander TL, van Mechelen W, Boot CRL, Roden LC, Lambert EV, Rae DE. Associations between self-reported sleep duration and mortality in employed individuals: systematic review and meta-analysis. Am J Health Promot. 2021;35(6):853–65. https://doi.org/10.1177/0890117121992288. Epub 11 Feb 2021. PMID: 33567861
39. Gallicchio L, Kalesan B. Sleep duration and mortality: a systematic review and meta-analysis. J Sleep Res. 2009;18(2):148–58. https://doi.org/10.1111/j.1365-2869.2008.00732.x. PMID: 19645960
40. García-Perdomo HA, Zapata-Copete J, Rojas-Cerón CA. Sleep duration and risk of all-cause mortality: a systematic review and meta-analysis. Epidemiol Psychiatr Sci. 2019;28(5):578–88. https://doi.org/10.1017/S2045796018000379. Epub 30 July 2018. PMID: 30058510; PMCID: PMC6998920
41. Mongrain V, Carrier J, Dumont M. Chronotype and sex effects on sleep architecture and quantitative sleep EEG in healthy young adults. Sleep. 2005;28(7):819–27. https://doi.org/10.1093/sleep/28.7.819. PMID: 16124660
42. von Gall C, Holub L, Pfeffer M, Eickhoff S. Chronotype-dependent sleep loss is associated with a lower amplitude in circadian rhythm and a higher fragmentation of REM sleep in young healthy adults. Brain Sci. 2023;13(10):1482. https://doi.org/10.3390/brainsci13101482. PMID: 37891848; PMCID: PMC10605513
43. Su F, Huang D, Wang H, Yang Z. Associations of shift work and night work with the risk of all-cause, cardiovascular and cancer mortality: a meta-analysis of cohort studies. Sleep Med. 2021;86:90–8. https://doi.org/10.1016/j.sleep.2021.08.017. Epub 21 Aug 2021. PMID: 34479052

44. Manderville RA, Wetmore SD. Mutagenicity of ochratoxin a: role for a carbon-linked C8-deoxyguanosine adduct? J Agric Food Chem. 2017;65(33):7097–105. https://doi.org/10.1021/acs.jafc.6b03897. Epub 15 Nov 2016. PMID: 28830149
45. Yaffe K, Falvey CM, Hoang T. Connections between sleep and cognition in older adults. Lancet Neurol. 2014;13(10):1017–28. https://doi.org/10.1016/S1474-4422(14)70172-3. PMID: 25231524
46. Brown MK, Chan MT, Zimmerman JE, Pack AI, Jackson NE, Naidoo N. Aging-induced endoplasmic reticulum stress alters sleep and sleep homeostasis. Neurobiol Aging. 2014;35(6):1431–41. https://doi.org/10.1016/j.neurobiolaging.201.
47. Helfrich RF, Mander BA, Jagust WJ, Knight RT, Walker MP. Old brains come uncoupled in sleep: slow wave-spindle synchrony, brain atrophy, and forgetting. Neuron. 2018;97(1):221–30.e4. https://doi.org/10.1016/j.neuron.2017.11.020. Epub 14 Dec 2017. PMID: 29249289; PMCID: PMC5754239
48. Rana BK, Panizzon MS, Franz CE, Spoon KM, Jacobson KC, Xian H, Ancoli-Israel S, Lyons M, Kremen WS. Association of sleep quality on memory-related executive functions in middle age. J Int Neuropsychol Soc. 2018;24(1):67–76. https://doi.org/10.1017/S1355617717000637. Epub 1 Aug 2017. PMID: 28760172; PMCID: PMC5958545
49. You Y, Chen Y, Liu R, Zhang Y, Wang M, Yang Z, Liu J, Ma X. Inverted U-shaped relationship between sleep duration and phenotypic age in US adults: a population-based study. Sci Rep. 2024;14(1):6247. https://doi.org/10.1038/s41598-024-56316-7. PMID: 38486063; PMCID: PMC10940593
50. Addo PNO, Mundagowa PT, Zhao L, Kanyangarara M, Brown MJ, Liu J. Associations between sleep duration, sleep disturbance, and cardiovascular disease biomarkers among adults in the United States. BMC Public Health. 2024;24(1):947. https://doi.org/10.1186/s12889-024-18381-5. PMID: 38566084; PMCID: PMC10985959
51. Horvath S. DNA methylation age of human tissues and cell types. Genome Biol. 2013;14(10):R115. https://doi.org/10.1186/gb-2013-14-10-r115. Erratum in: Genome Biol. 2015;16:96. https://doi.org/10.1186/s13059-015-0649-6. PMID: 24138928; PMCID: PMC4015143.
52. Horvath S, Ritz BR. Increased epigenetic age and granulocyte counts in the blood of Parkinson's disease patients. Aging (Albany NY). 2015;7(12):1130–42. https://doi.org/10.18632/aging.100859. PMID: 26655927; PMCID: PMC4712337
53. Horvath S, Zhang Y, Langfelder P, Kahn RS, Boks MP, van Eijk K, van den Berg LH, Ophoff RA. Aging effects on DNA methylation modules in human brain and blood tissue. Genome Biol. 2012;13(10):R97. https://doi.org/10.1186/gb-2012-13-10-r97. PMID: 23034122; PMCID: PMC4053733
54. Giannakou ME, Goss M, Jünger MA, Hafen E, Leevers SJ, Partridge L. Long-lived Drosophila with overexpressed dFOXO in an adult fat body. Science. 2004;305(5682):361. https://doi.org/10.1126/science.1098219. Epub 10 June 2004. PMID: 15192154
55. Blüher M, Kahn BB, Kahn CR. Extended longevity in mice lacking the insulin receptor in adipose tissue. Science. 2003;299(5606):572–4. https://doi.org/10.1126/science.1078223. PMID: 12543978
56. Tan Q, Soerensen M, Kruse TA, Christensen K, Christiansen L. A novel permutation test for case-only analysis identifies epistatic effects on human longevity in the FOXO gene family. Aging Cell. 2013;12(4):690–4. https://doi.org/10.1111/acel.12092. Epub 15 May 2013. PMID: 23607278; PMCID: PMC3714332
57. Peters MJ, Joehanes R, Pilling LC, Schurmann C, Conneely KN, Powell J, Reinmaa E, Sutphin GL, Zhernakova A, Schramm K, Wilson YA, Kobes S, Tukiainen T, NABEC/UKBEC Consortium, Ramos YF, Göring HH, Fornage M, Liu Y, Gharib SA, Stranger BE, De Jager PL, Aviv A, Levy D, Murabito JM, Munson PJ, Huan T, Hofman A, Uitterlinden AG, Rivadeneira F, van Rooij J, Stolk L, Broer L, Verbiest MM, Jhamai M, Arp P, Metspalu A, Tserel L, Milani L, Samani NJ, Peterson P, Kasela S, Codd V, Peters A, Ward-Caviness CK, Herder C, Waldenberger M, Roden M, Singmann P, Zeilinger S, Illig T, Homuth G, Grabe

HJ, Völzke H, Steil L, Kocher T, Murray A, Melzer D, Yaghootkar H, Bandinelli S, Moses EK, Kent JW, Curran JE, Johnson MP, Williams-Blangero S, Westra HJ, McRae AF, Smith JA, Kardia SL, Hovatta I, Perola M, Ripatti S, Salomaa V, Henders AK, Martin NG, Smith AK, Mehta D, Binder EB, Nylocks KM, Kennedy EM, Klengel T, Ding J, Suchy-Dicey AM, Enquobahrie DA, Brody J, Rotter JI, Chen YD, Houwing-Duistermaat J, Kloppenburg M, Slagboom PE, Helmer Q, den Hollander W, Bean S, Raj T, Bakhshi N, Wang QP, Oyston LJ, Psaty BM, Tracy RP, Montgomery GW, Turner ST, Blangero J, Meulenbelt I, Ressler KJ, Yang J, Franke L, Kettunen J, Visscher PM, Neely GG, Korstanje R, Hanson RL, Prokisch H, Ferrucci L, Esko T, Teumer A, van Meurs JB, Johnson AD. The transcriptional landscape of age in human peripheral blood. Nat Commun. 2015;6:8570. https://doi.org/10.1038/ncomms9570. PMID: 26490707; PMCID: PMC4639797

58. Westerterp M, Tsuchiya K, Tattersall IW, Fotakis P, Bochem AE, Molusky MM, Ntonga V, Abramowicz S, Parks JS, Welch CL, Kitajewski J, Accili D, Tall AR. Deficiency of ATP-binding cassette transporters A1 and G1 in endothelial cells accelerates atherosclerosis in mice. Arterioscler Thromb Vasc Biol. 2016;36(7):1328–37. https://doi.org/10.1161/ATVBAHA.115.306670. Epub 19 May 2016. PMID: 27199450; PMCID: PMC4919153
59. D'Aquila P, Crocco P, De Rango F, Indiveri C, Bellizzi D, Rose G, Passarino G. A genetic variant of ASCT2 hampers in vitro RNA splicing and correlates with human longevity. Rejuvenation Res. 2018;21(3):193–9. https://doi.org/10.1089/rej.2017.1948. Epub 8 Sept 2017. PMID: 28885889
60. Nakamura S, Kawai K, Takeshita Y, Honda M, Takamura T, Kaneko S, Matoba R, Matsubara K. Identification of blood biomarkers of aging by transcript profiling of whole blood. Biochem Biophys Res Commun. 2012;418(2):313–8. https://doi.org/10.1016/j.bbrc.2012.01.018. Epub 12 Jan 2012. PMID: 22266314
61. Huan T, Chen G, Liu C, Bhattacharya A, Rong J, Chen BH, Seshadri S, Tanriverdi K, Freedman JE, Larson MG, Murabito JM, Levy D. Age-associated microRNA expression in human peripheral blood is associated with all-cause mortality and age-related traits. Aging Cell. 2018;17(1):e12687. https://doi.org/10.1111/acel.12687. Epub 17 Oct 2017. PMID: 29044988; PMCID: PMC5770777
62. Kinser HE, Pincus Z. MicroRNAs as modulators of longevity and the aging process. Hum Genet. 2020;139(3):291–308. https://doi.org/10.1007/s00439-019-02046-0. Epub 11 July 2019. PMID: 31297598; PMCID: PMC6954352
63. Pereira Fernandes D, Bitar M, Jacobs FMJ, Barry G. Long non-coding RNAs in neuronal aging. Noncoding RNA. 2018;4(2):12. https://doi.org/10.3390/ncrna4020012. PMID: 29670042; PMCID: PMC6027360
64. Johnson AA, Shokhirev MN, Wyss-Coray T, Lehallier B. Systematic review and analysis of human proteomics aging studies unveils a novel proteomic aging clock and identifies key processes that change with age. Ageing Res Rev. 2020;60:101070. https://doi.org/10.1016/j.arr.2020.101070. Epub 18 Apr 2020. PMID: 32311500
65. Tanaka T, Biancotto A, Moaddel R, Moore AZ, Gonzalez-Freire M, Aon MA, Candia J, Zhang P, Cheung F, Fantoni G, CHI Consortium, Semba RD, Ferrucci L. Plasma proteomic signature of age in healthy humans. Aging Cell. 2018;17(5):e12799. https://doi.org/10.1111/acel.12799. Epub 11 July 2018. PMID: 29992704; PMCID: PMC6156492
66. Pastushkova LK, Kononikhin AS, Tiys ES, Dobrokhotov IV, Ivanisenko VA, Nikolaev EN, Larina IM, Popov I. Characteristics of age-dependent changes in urine proteome in healthy men. Adv Gerontol. 2016;6:123–8.
67. Santos-Lozano A, Valenzuela PL, Llavero F, Lista S, Carrera-Bastos P, Hampel H, Pareja-Galeano H, Gálvez BG, López JA, Vázquez J, Emanuele E, Zugaza JL, Lucia A. Successful aging: insights from proteome analyses of healthy centenarians. Aging (Albany NY). 2020;12(4):3502–15. https://doi.org/10.18632/aging.102826. Epub 25 Feb 2020. PMID: 32100723; PMCID: PMC7066932
68. Basisty N, Kale A, Jeon OH, Kuehnemann C, Payne T, Rao C, Holtz A, Shah S, Sharma V, Ferrucci L, Campisi J, Schilling B. A proteomic atlas of senescence-associated secretomes

for aging biomarker development. PLoS Biol. 2020;18(1):e3000599. https://doi.org/10.1371/journal.pbio.3000599. PMID: 31945054; PMCID: PMC6964821
69. Vanhooren V, Navarrete Santos A, Voutetakis K, Petropoulos I, Libert C, Simm A, Gonos ES, Friguet B. Protein modification and maintenance systems as biomarkers of aging. Mech Ageing Dev. 2015;151:71–84. https://doi.org/10.1016/j.mad.2015.03.009. Epub 3 Apr 2015. PMID: 25846863
70. Robinson O, Chadeau Hyam M, Karaman I, Climaco Pinto R, Ala-Korpela M, Handakas E, Fiorito G, Gao H, Heard A, Jarvelin MR, Lewis M, Pazoki R, Polidoro S, Tzoulaki I, Wielscher M, Elliott P, Vineis P. Determinants of accelerated metabolomic and epigenetic aging in a UK cohort. Aging Cell. 2020;19(6):e13149. https://doi.org/10.1111/acel.13149. Epub 3 May 2020. PMID: 32363781; PMCID: PMC7294785
71. Jové M, Maté I, Naudí A, Mota-Martorell N, Portero-Otín M, De la Fuente M, Pamplona R. Human aging is a metabolome-related matter of gender. J Gerontol A Biol Sci Med Sci. 2016;71(5):578–85. https://doi.org/10.1093/gerona/glv074. Epub 26 May 2015. PMID: 26019184
72. Ohayon MM, Carskadon MA, Guilleminault C, Vitiello MV. Meta-analysis of quantitative sleep parameters from childhood to old age in healthy individuals: developing normative sleep values across the human lifespan. Sleep. 2004;27(7):1255–73. https://doi.org/10.1093/sleep/27.7.1255. PMID: 15586779
73. Klerman EB, Dijk DJ. Age-related reduction in the maximal capacity for sleep—implications for insomnia. Curr Biol. 2008;18(15):1118–23. https://doi.org/10.1016/j.cub.2008.06.047. Epub 24 July 2008. PMID: 18656358; PMCID: PMC2582347
74. Frank MG. Sleep and developmental plasticity not just for kids. Prog Brain Res. 2011;193:221–32. https://doi.org/10.1016/B978-0-444-53839-0.00014-4. PMID: 21854965
75. Feinberg I, Campbell IG. Sleep EEG changes during adolescence: an index of a fundamental brain reorganization. Brain Cogn. 2010;72(1):56–65. https://doi.org/10.1016/j.bandc.2009.09.008. Epub 2 Nov 2009. PMID: 19883968
76. Larsen LH, Moe KE, Vitiello MV, Prinz PN. Age trends in the sleep EEG of healthy older men and women. J Sleep Res. 1995;4(3):160–72. https://doi.org/10.1111/j.1365-2869.1995.tb00165.x. PMID: 10607155
77. Phillips AJ, Chen PY, Robinson PA. Probing the mechanisms of chronotype using quantitative modeling. J Biol Rhythms. 2010;25(3):217–27. https://doi.org/10.1177/0748730410369208. PMID: 20484693
78. Gursoy AY, Kiseli M, Caglar GS. Melatonin in aging women. Climacteric. 2015;18(6):790–6. https://doi.org/10.3109/13697137.2015.1052393. Epub 25 Sept 2015. PMID: 26029988
79. Mander BA, Zhu AH, Lindquist JR, Villeneuve S, Rao V, Lu B, Saletin JM, Ancoli-Israel S, Jagust WJ, Walker MP. White matter structure in older adults moderates the benefit of sleep spindles on motor memory consolidation. J Neurosci. 2017;37(48):11675–87. https://doi.org/10.1523/JNEUROSCI.3033-16.2017. Epub 30 Oct 2017. PMID: 29084867; PMCID: PMC5707766
80. Besedovsky L, Lange T, Haack M. The sleep-immune crosstalk in health and disease. Physiol Rev. 2019;99(3):1325–80. https://doi.org/10.1152/physrev.00010.2018. PMID: 30920354; PMCID: PMC6689741
81. Lin WC, Huang CC, Chen HL, Chou KH, Chen PC, Tsai NW, Chen MH, Friedman M, Lin HC, Lu CH. Longitudinal brain structural alterations and systemic inflammation in obstructive sleep apnea before and after surgical treatment. J Transl Med. 2016;14(1):139. https://doi.org/10.1186/s12967-016-0887-8. PMID: 27188598; PMCID: PMC4901987
82. Gouveris H, Bahr K, Jahn C, Matthias C, Simon P. The apnea-hypopnea index underestimates systemic inflammation in women with sleep-disordered breathing. J Womens Health (Larchmt). 2018;27(7):920–6. https://doi.org/10.1089/jwh.2017.6819. Epub 9 Apr 2018. PMID: 29630436

83. Chen L, Deng H, Cui H, Fang J, Zuo Z, Deng J, Li Y, Wang X, Zhao L. Inflammatory responses and inflammation-associated diseases in organs. Oncotarget. 2017;9(6):7204–18. https://doi.org/10.18632/oncotarget.23208. PMID: 29467962; PMCID: PMC5805548
84. Dantzer R, O'Connor JC, Freund GG, Johnson RW, Kelley KW. From inflammation to sickness and depression: when the immune system subjugates the brain. Nat Rev Neurosci. 2008;9(1):46–56. https://doi.org/10.1038/nrn2297. PMID: 18073775; PMCID: PMC2919277
85. Kapsimalis F, Richardson G, Opp MR, Kryger M. Cytokines and normal sleep. Curr Opin Pulm Med. 2005;11(6):481–4. https://doi.org/10.1097/01.mcp.0000183062.98665.6b. PMID: 16217172
86. Ibarra-Coronado EG, Pantaleón-Martínez AM, Velazquéz-Moctezuma J, Prospéro-García O, Méndez-Díaz M, Pérez-Tapia M, Pavón L, Morales-Montor J. The bidirectional relationship between sleep and immunity against infections. J Immunol Res. 2015;2015:678164. https://doi.org/10.1155/2015/678164. Epub 31 Aug 2015. PMID: 26417606; PMCID: PMC4568388
87. Fernandez-Mendoza J, Baker JH, Vgontzas AN, Gaines J, Liao D, Bixler EO. Insomnia symptoms with objective short sleep duration are associated with systemic inflammation in adolescents. Brain Behav Immun. 2017;61:110–6. https://doi.org/10.1016/j.bbi.2016.12.026. Epub 30 Dec 2016. PMID: 28041986; PMCID: PMC5316336
88. Carroll JE, Carrillo C, Olmstead R, Witarama T, Breen EC, Yokomizo M, Seeman T, Irwin MR. Sleep deprivation and divergent toll-like receptor-4 activation of cellular inflammation in aging. Sleep. 2015;38(2):205–11. https://doi.org/10.5665/sleep.4398. PMID: 25325509; PMCID: PMC4288601
89. Slavish DC, Graham-Engeland JE, Engeland CG, Taylor DJ, Buxton OM. Insomnia symptoms are associated with elevated C-reactive protein in young adults. Psychol Health. 2018;33(11):1396–415. https://doi.org/10.1080/08870446.2018.1500577. Epub 25 Oct 2018. PMID: 30358412
90. Davis KC, Raizen DM. A mechanism for sickness sleep: lessons from invertebrates. J Physiol. 2017;595(16):5415–24. https://doi.org/10.1113/JP273009. Epub 22 Feb 2017. PMID: 28028818; PMCID: PMC5556163
91. Hill AJ, Mansfield R, Lopez JM, Raizen DM, Van Buskirk C. Cellular stress induces a protective sleep-like state in C. elegans. Curr Biol. 2014;24(20):2399–405. https://doi.org/10.1016/j.cub.2014.08.040. Epub 25 Sept 2014. PMID: 25264259; PMCID: PMC4254280
92. Lenz O, Xiong J, Nelson MD, Raizen DM, Williams JA. FMRFamide signaling promotes stress-induced sleep in Drosophila. Brain Behav Immun. 2015;47:141–8. https://doi.org/10.1016/j.bbi.2014.12.028. Epub 7 Feb 2015. PMID: 25668617; PMCID: PMC4467992
93. Lange T, Dimitrov S, Bollinger T, Diekelmann S, Born J. Sleep after vaccination boosts immunological memory. J Immunol. 2011;187(1):283–90. https://doi.org/10.4049/jimmunol.1100015. Epub 1 June 2011. PMID: 21632713
94. Opp MR. Sleeping to fuel the immune system: mammalian sleep and resistance to parasites. BMC Evol Biol. 2009;9:8. https://doi.org/10.1186/1471-2148-9-8. PMID: 19134176; PMCID: PMC2633283
95. Fernandes ER, Barbosa ML, Amaral MP, de Souza Apostolico J, Sulczewski FB, Tufik S, Andersen ML, Boscardin SB, Keller AC, Rosa DS. Sleep disturbance during infection compromises Tfh differentiation and impacts host immunity. iScience. 2020;23(10):101599. https://doi.org/10.1016/j.isci.2020.101599. PMID: 33205014; PMCID: PMC7648138
96. Orzech KM, Acebo C, Seifer R, Barker D, Carskadon MA. Sleep patterns are associated with common illnesses in adolescents. J Sleep Res. 2014;23(2):133–42. https://doi.org/10.1111/jsr.12096. Epub 18 Oct 2013. PMID: 24134661; PMCID: PMC4115328
97. Palma JA, Urrestarazu E, Iriarte J. Sleep loss as risk factor for neurologic disorders: a review. Sleep Med. 2013;14(3):229–36. https://doi.org/10.1016/j.sleep.2012.11.019. Epub 23 Jan 2013. PMID: 23352029
98. Sabia S, Fayosse A, Dumurgier J, van Hees VT, Paquet C, Sommerlad A, Kivimäki M, Dugravot A, Singh-Manoux A. Association of sleep duration in middle and old age with incidence of dementia. Nat Commun. 2021;12(1):2289. https://doi.org/10.1038/s41467-021-22354-2. PMID: 33879784; PMCID: PMC8058039

99. Sadeghmousavi S, Eskian M, Rahmani F, Rezaei N. The effect of insomnia on the development of Alzheimer's disease. J Neuroinflamm. 2020;17(1):289. https://doi.org/10.1186/s12974-020-01960-9. PMID: 33023629; PMCID: PMC7542374
100. Hsiao YH, Chen YT, Tseng CM, Wu LA, Lin WC, Su VY, Perng DW, Chang SC, Chen YM, Chen TJ, Lee YC, Chou KT. Sleep disorders and increased risk of autoimmune diseases in individuals without sleep apnea. Sleep. 2015;38(4):581–6. https://doi.org/10.5665/sleep.4574. PMID: 25669189; PMCID: PMC4355897
101. Leng Y, Cappuccio FP, Wainwright NW, Surtees PG, Luben R, Brayne C, Khaw KT. Sleep duration and risk of fatal and nonfatal stroke: a prospective study and meta-analysis. Neurology. 2015;84(11):1072–9. https://doi.org/10.1212/WNL.0000000000001371. Epub 25 Feb 2015. PMID: 25716357; PMCID: PMC4371404
102. Cappuccio FP, Cooper D, D'Elia L, Strazzullo P, Miller MA. Sleep duration predicts cardiovascular outcomes: a systematic review and meta-analysis of prospective studies. Eur Heart J. 2011;32(12):1484–92. https://doi.org/10.1093/eurheartj/ehr007. Epub 7 Feb 2011. PMID: 21300732
103. Domínguez F, Fuster V, Fernández-Alvira JM, Fernández-Friera L, López-Melgar B, Blanco-Rojo R, Fernández-Ortiz A, García-Pavía P, Sanz J, Mendiguren JM, Ibañez B, Bueno H, Lara-Pezzi E, Ordovás JM. Association of sleep duration and quality with subclinical atherosclerosis. J Am Coll Cardiol. 2019;73(2):134–44. https://doi.org/10.1016/j.jacc.2018.10.060. PMID: 30654884
104. Kakizaki M, Kuriyama S, Sone T, Ohmori-Matsuda K, Hozawa A, Nakaya N, Fukudo S, Tsuji I. Sleep duration and the risk of breast cancer: the Ohsaki Cohort Study. Br J Cancer. 2008;99(9):1502–5. https://doi.org/10.1038/sj.bjc.6604684. Epub 23 Sept 2008. PMID: 18813313; PMCID: PMC2579702
105. Kakizaki M, Inoue K, Kuriyama S, Sone T, Matsuda-Ohmori K, Nakaya N, Fukudo S, Tsuji I, Ohsaki Cohort Study. Sleep duration and the risk of prostate cancer: the Ohsaki Cohort Study. Br J Cancer. 2008;99(1):176–8. https://doi.org/10.1038/sj.bjc.6604425. Epub 10 June 2008. PMID: 18542076; PMCID: PMC2453016
106. Jiao L, Duan Z, Sangi-Haghpeykar H, Hale L, White DL, El-Serag HB. Sleep duration and incidence of colorectal cancer in postmenopausal women. Br J Cancer. 2013;108(1):213–21. https://doi.org/10.1038/bjc.2012.561. Epub 3 Jan 2013. PMID: 23287986; PMCID: PMC3553538
107. Cao J, Eshak ES, Liu K, Muraki I, Cui R, Iso H, Tamakoshi A, JACC Study Group. Sleep duration and risk of breast cancer: The JACC Study. Breast Cancer Res Treat. 2019;174(1):219–25. https://doi.org/10.1007/s10549-018-4995-4. Epub 20 Nov 2018. PMID: 30460465
108. Wu AH, Wang R, Koh WP, Stanczyk FZ, Lee HP, Yu MC. Sleep duration, melatonin and breast cancer among Chinese women in Singapore. Carcinogenesis. 2008;29(6):1244–8. https://doi.org/10.1093/carcin/bgn100. Epub 30 Apr 2008. PMID: 18448486; PMCID: PMC2574570
109. Wehr TA. The duration of human melatonin secretion and sleep respond to changes in day length (photoperiod). J Clin Endocrinol Metab. 1991;73(6):1276–80. https://doi.org/10.1210/jcem-73-6-1276. PMID: 1955509
110. Tan BKJ, Tan NKW, Teo YH, Yap DWT, Raghupathy J, Gao EY, Toh ST, See A. Association of obstructive sleep apnea with thyroid cancer incidence: a systematic review and meta-analysis. Eur Arch Otorhinolaryngol. 2022;279(11):5407–14. https://doi.org/10.1007/s00405-022-07457-w. Epub 16 June 2022. PMID: 35708764
111. Cheng L, Guo H, Zhang Z, Yao Y, Yao Q. Obstructive sleep apnea and incidence of malignant tumors: a meta-analysis. Sleep Med. 2021;84:195–204. https://doi.org/10.1016/j.sleep.2021.05.029. Epub 21 June 2021. PMID: 34166986
112. Mashaqi S, Gozal D. The impact of obstructive sleep apnea and PAP therapy on all-cause and cardiovascular mortality based on age and gender—a literature review. Respir Investig. 2020;58(1):7–20. https://doi.org/10.1016/j.resinv.2019.08.002. Epub 18 Oct 2019. PMID: 31631059

113. Li L, Gan Y, Zhou X, Jiang H, Zhao Y, Tian Q, He Y, Liu Q, Mei Q, Wu C, Lu Z. Insomnia and the risk of hypertension: A meta-analysis of prospective cohort studies. Sleep Med Rev. 2021;56:101403. https://doi.org/10.1016/j.smrv.2020.101403. Epub 28 Nov 2020. PMID: 33360604
114. Zhang Y, Jiang X, Liu J, Lang Y, Liu Y. The association between insomnia and the risk of metabolic syndrome: A systematic review and meta-analysis. J Clin Neurosci. 2021;89:430–6. https://doi.org/10.1016/j.jocn.2021.05.039. Epub 29 May 2021. Erratum in: J Clin Neurosci. 2021;93:287. https://doi.org/10.1016/j.jocn.2021.08.021. PMID: 34074598.
115. Hu S, Lan T, Wang Y, Ren L. Individual insomnia symptom and increased hazard risk of cardiocerebral vascular diseases: a meta-analysis. Front Psychiatry. 2021;12:654719. https://doi.org/10.3389/fpsyt.2021.654719. PMID: 34054612; PMCID: PMC8160242
116. Zhang Y, Zhu Y, Zhang X, Li C, Fu H, Lin L, Yang Z, Zhang B. The association of sleep duration and leukocyte telomere length in middle-aged and young-old adults: A cross-sectional study of UK Biobank. Sleep Med. 2024;117:18–24. https://doi.org/10.1016/j.sleep.2024.02.043. Epub 7 Mar 2024. PMID: 38493659

GPSR Compliance

The European Union's (EU) General Product Safety Regulation (GPSR) is a set of rules that requires consumer products to be safe and our obligations to ensure this.

If you have any concerns about our products, you can contact us on ProductSafety@springernature.com

In case Publisher is established outside the EU, the EU authorized representative is:

Springer Nature Customer Service Center GmbH
Europaplatz 3
69115 Heidelberg, Germany

Batch number: 10755216

Printed by Printforce, the Netherlands